OPERATING MANUAL FOR SPACESHIP EARTH

R. Buckminster Fuller (1895–1983) was an American architect, engineer, inventor, philosopher, and poet who established a reputation as one of the most original thinkers of the second half of the twentieth century. In 1927 he resolved to devote his life to a nonprofit search for design patterns that could maximize the social uses of the world's energy resources and evolving industrial complex. "I started with the universe—as an organization of energy systems of which all our experiences and possible experiences are only local instances." Fuller intended that all his developments contribute to the growth of a global strategy for solving world problems by finding ways to do more with less. His best-known inventions were the geodesic dome and the Dymaxion car, but he also patented twenty-four other inventions and was honored with dozens of international awards and honorary degrees. On the recommendation of the Royal Institute of British Architects, Queen Elizabeth II awarded Fuller the Royal Gold Medal for Architecture. In 1968 he received the Gold Medal Award of the National Institute of Arts and Letters, His books include *Nine Chains to The Moon* (1938); *Education Automatiion: Freeing the Scholar to Return to His Studies* (1962); *No More Secondhand God, and Other Writings* (1962); *Ideas and Integrities: A Spontaneous Autobiographical Disclosure* (1963); *Utopia or Oblivion: The Prospects for Humanity* (1969); and *Operating Manual For Spaceship Earth* (1969).

R. Buckminster Fuller

Operating Manual for Spaceship Earth

ARKANA

ARKANA
Published by the Penguin Group
Penguin Books USA Inc., 375 Hudson Street,
New York, New York 10014, U.S.A.
Penguin Books Ltd, 27 Wrights Lane,
London W8 5TZ, England
Penguin Books Australia Ltd, Ringwood,
Victoria, Australia
Penguin Books Canada Ltd, 10 Alcorn Avenue,
Toronto, Ontario, Canada M4V 3B2
Penguin Books (N.Z.) Ltd, 182–190 Wairau Road,
Auckland 10, New Zealand

Penguin Books Ltd, Registered Offices:
Harmondsworth, Middlesex, England

Paperback edition first published in the United States of America by E. P. Dutton
Published simultaneously in Canada by Fitzhenry and Whiteside, Limited, Toronto
Published in Arkana 1991

10 9 8 7 6 5 4 3

ISBN 0 14 01.9451 7

Printed in the United States of America

contents

operating manual
for
spaceship EARTH

1 *comprehensive propensities*

I AM enthusiastic over humanity's extraordinary and sometimes very timely ingenuities. If you are in a shipwreck and all the boats are gone, a piano top buoyant enough to keep you afloat that comes along makes a fortuitous life preserver. But this is not to say that the best way to design a life preserver is in the form of a piano top. I think that we are clinging to a great many piano tops in accepting yesterday's fortuitous contrivings as constituting the only means for solving a given problem. Our brains deal exclusively with special-case experiences.

Only our minds are able to discover the generalized principles operating without exception in each and every special-experience case which if detected and mastered will give knowledgeable advantage in all instances.

Because our spontaneous initiative has been frustrated, too often inadvertently, in earliest childhood we do not tend, customarily, to dare to think competently regarding our potentials. We find it socially easier to go on with our narrow, shortsighted specializations and leave it to others—primarily to the politicians—to find some way of resolving our common dilemmas. Countering that spontaneous grown-up trend to narrowness I will do my, hopefully "childish," best to confront as many of our problems as possible by employing the longest-distance thinking of which I am capable—though that may not take us very far into the future.

Having been trained at the U. S. Naval Academy and practically experienced in the powerfully effective forecasting arts of celestial navigation, pilotage, ballistics, and logistics, and in the long-range, anticipatory, design science governing yesterday's naval mastery of the world from which our present day's general systems theory has been derived, I recall that in 1927 I set about deliberately exploring to see how far ahead we could make competent forecasts regarding the direction in which all humanity is trending and to see how effectively we could interpret the physical details of what comprehensive evolution might be portending as disclosed by the available data. I came to the conclusion that it is possible to make a fairly reasonable forecast of about twenty-five years. That seems to be about one industrial "tooling" generation. On the average, all inventions seem to get melted up about every twenty-five years, after which the metals come back into recirculation in new and usually more effective uses. At any rate, in 1927 I evolved a forecast. Most of my 1927's prognosticating went only to 1952—that is, for a quarter-century, but some of it went on for a half-century, to 1977.

In 1927 when people had occasion to ask me about my prognostications and I told them what I thought it would be appropriate to do about what I could see ahead for the 1950's, 1960's, and 1970's people used to say to me, "Very amusing—you are a thousand years ahead of your time." Having myself studied the increments in which we can think forwardly I was amazed at the ease with which the rest of society seemed to be able to see a thousand years ahead while I could see only one-fortieth of that time distance. As time went on people began to tell me that I was a hundred years ahead, and now they tell me that I'm a little behind the times. But I have learned about public reaction to the unfamiliar and also about the ease and speed with which the transformed reality becomes so "natural" as misseemingly to have been always obvious. So I knew that their last observations were made only because the evolutionary events I had foreseen have occurred on schedule.

However, all that experience gives me confidence in discussing the next quarter-century's events. First, I'd like to explore a few thoughts about the vital data confronting us right now—such as the fact that more than half of humanity as yet exists in miserable poverty, prematurely doomed, unless we alter our comprehensive physical circumstances. It is certainly no solution to evict the poor, replacing their squalid housing with much more expensive buildings which the original tenants can't afford to reoccupy. Our society adopts many such superficial palliatives. Because yesterday's negatives are moved out of sight from their familiar locations many persons are willing to pretend to themselves that the problems have been solved. I feel that one of the reasons why we are struggling inadequately today is that we reckon our costs on too shortsighted a basis and are later overwhelmed with the unexpected costs brought about by our shortsightedness.

Of course, our failures are a consequence of many factors, but possibly one of the most important is the fact that society operates on the theory that specialization is the key to success, not realizing that specialization precludes comprehensive thinking. This means that the potentially-integratable—techno-economic advantages accruing to society from the myriad specializations are not comprehended integratively and therefore are not realized, or they are realized only in negative ways, in new weaponry or the industrial support only of warfaring.

All universities have been progressively organized for ever finer specialization. Society assumes that specialization is natural, inevitable, and desirable. Yet in observing a little child, we find it is interested in everything and spontaneously apprehends, comprehends, and co-ordinates an ever-expanding inventory of experiences. Children are enthusiastic planetarium audiences. Nothing seems to be more prominent about human life than its wanting to understand all and put everything together.

One of humanity's prime drives is to understand and be understood. All other living creatures are designed for highly specialized tasks. Man seems unique as the comprehensive comprehender and co-ordinator of local universe affairs. If the total scheme of nature required man to be a specialist she would have made him so by having him born with one eye and a microscope attached to it.

What nature needed man to be was adaptive in many if not any direction; wherefore she gave man a mind as well as a co-ordinating switchboard brain. Mind apprehends and comprehends the general principles governing flight and deep sea diving, and man puts on his wings or his lungs, then takes them off when not using them. The specialist bird is greatly impeded by its wings when trying to walk. The fish cannot come out of the sea and walk upon land, for birds and fish are specialists.

Of course, we are beginning to learn a little in the behavioral sciences regarding how little we know about children and the educational processes. We had assumed the child to be an empty brain receptacle into which we could inject our methodically-gained wisdom until that child, too, became educated. In the light of modern behavioral science experiments that was not a good working assumption.

Inasmuch as the new life always manifests comprehensive propensities I would like to know why it is that we have disregarded all children's significantly spontaneous and comprehensive curiosity and in our formal education have deliberately instituted processes leading only to narrow specialization. We do not have to go very far back in history for the answer. We get back to great, powerful men of the sword, exploiting their prowess fortuitously and ambitiously, surrounded by the abysmal ignorance of world society. We find early society struggling under economic conditions wherein less than 1 per cent of humanity seemed able to live its full span of years. This forlorn economic prospect resulted from the seeming inadequacy of vital resources and from an illiterate society's inability to cope successfully with the environment, while saddled also with preconditioned instincts which inadvertently produced many new human babies. Amongst the strugglers we had cunning leaders who said, "Follow me, and we'll make out better than the others." It was the most powerful and shrewd of these leaders who, as we shall see, invented and developed specialization.

Looking at the total historical pattern of man around the Earth and observing that three-quarters of the Earth is water, it seems obvious why men, unaware that they would some day contrive to fly and penetrate the ocean in submarines, thought of themselves exclusively as pedestrians—as dry land specialists. Confined to the quarter of the Earth's surface which is dry land it is easy to see how they came to specialize further as farmers or hunters—or, commanded by their leader, became specialized as soldiers. Less than half of the dry 25 per cent of the Earth's surface was immediately favorable to the support of human life. Thus, throughout history 99.9 per cent of humanity has occupied only 10 per cent of the total Earth surface, dwelling only where life support was visibly obvious. The favorable land was not in one piece, but consisted of a myriad of relatively small parcels widely dispersed over the surface of the enormous Earth sphere. The small isolated groups of humanity were utterly unaware of one another's existence. They were everywhere ignorant of the vast variety of very different environments and resource patterns occurring other than where they dwelt.

But there were a few human beings who gradually, through the process of invention and experiment, built and operated, first, local river and bay, next, along-shore, then off-shore rafts, dugouts, grass broats, and outrigger sailing canoes. Finally, they developed voluminous rib-bellied fishing vessels, and thereby ventured out to sea for progressively longer periods. Developing ever larger and more capable ships, the seafarers eventually were able to remain for months on the high seas. Thus, these venturers came to live normally at sea. This led them inevitably into world-around, swift, fortune-producing enterprise. Thus they became the first world men.

The men who were able to establish themselves on the oceans had also to be extraordinarily effective with the sword upon both land and sea. They had also to have great anticipatory vision, great ship designing capability, and original scientific conceptioning, mathematical skill in navigation and exploration techniques for coping in fog, night, and storm with the invisible hazards of rocks, shoals, and currents. The great sea venturers had to be able to command all the people in their dry land realm in order to commandeer the adequate metalworking, woodworking, weaving, and other skills necessary to produce their large, complex ships. They had to establish and maintain their authority in order that they themselves and the craftsmen preoccupied in producing the ship be adequately fed by the food-producing hunters and farmers of their realm. Here we see the specialization being greatly amplified under the supreme authority of the comprehensively visionary and brilliantly co-ordinated top swordsman, sea venturer. If his "ship came in" —that is, returned safely from its years' long venturing—all the people in his realm prospered and their leader's power was vastly amplified.

There were very few of these top power

men. But as they went on their sea ventures they gradually found that the waters interconnected all the world's people and lands. They learned this unbeknownst to their illiterate sailors, who, often as not, having been hit over the head in a saloon and dragged aboard to wake up at sea, saw only a lot of water and, without navigational knowledge, had no idea where they had travelled.

The sea masters soon found that the people in each of the different places visited knew nothing of people in other places. The great venturers found the resources of Earth very unevenly distributed, and discovered that by bringing together various resources occurring remotely from one another one complemented the other in producing tools, services, and consumables of high advantage and value. Thus resources in one place which previously had seemed to be absolutely worthless suddenly became highly valued. Enormous wealth was generated by what the sea venturers could do in the way of integrating resources and distributing the products to the, everywhere around the world, amazed and eager customers. The ship-owning captains found that they could carry fantastically large cargoes in their ships, due to nature's floatability—cargoes so large they could not possibly be carried on the backs of animals or the backs of men. Furthermore, the ships could sail across a bay or sea, travelling shorter distances in much less time than it took to go around the shores and over the intervening mountains. So these very few masters of the water world beçame incalculably rich and powerful.

To understand the development of *intellectual specialization,* which is our first objective, we must study further the comprehensive intellectual capabilities of the sea leaders in contradistinction to the myriad of physical, muscle, and craft-skill specializations which their intellect and their skillful swordplay commanded. The great sea venturers thought always in terms of the world, because the world's waters are continuous and cover three-quarters of the Earth planet. This meant that before the invention and use of cables and wireless 99.9 per cent of humanity thought only in the terms of their own local terrain. Despite our recently developed communications intimacy and popular awareness of total Earth we, too, in 1969 are as yet politically organized entirely in the terms of exclusive and utterly obsolete sovereign separateness.

This "sovereign"—meaning top-weapons enforced—"national" claim upon humans born in various lands leads to ever more severely specialized servitude and highly personalized identity classification. As a consequence of the slavish "categoryitis" the scientifically illogical, and as we shall see, often meaningless questions "Where do you live?" "What are you?" "What religion?" "What race?" "What nationality?" are all thought of today as logical questions. By the twenty-first century it either will have become evident to humanity that these questions are absurd and anti-evolutionary or men will no longer be living on Earth. If you don't comprehend why that is so, listen to me closely.

2 *origins of specialization*

OBVIOUSLY we need to pursue further the origins of specialization into deep history, hoping thereby to correct or eliminate our erroneous concepts. Historically we can say that average human beings throughout pre-twentieth-century history had each seen only about one-millionth of the surface of their spherical Earth. This limited experience gave humans a locally-focused, specialized viewpoint. Not surprisingly, humanity thought the world was flat, and not surprisingly humans thought its horizontally-extended plane went circularly outward to infinity. In our schools today we still start off the education of our children by giving them planes and lines that go on, incomprehendibly "forever" toward a meaningless infinity. Such oversimplified viewpoints are misleading, blinding, and debilitating, because they preclude possible discovery of the significance of our integrated experiences.

Under these everyday, knowledge-thwarting or limiting circumstances of humanity, the comprehensively-informed master venturers of history who went to sea soon realized that the only real competition they had was that of other powerful outlaws who might also know or hope to learn through experience "what it is *all* about." I call these sea mastering people the *great outlaws* or *Great Pirates*—the G. P.'s—simply because the arbitrary laws enacted or edicted by men on the land could not be extended effectively to control humans beyond their shores and out upon the seas. So the world men who lived on the seas were inherently outlaws, and the only laws that could and did rule them were the natural laws—the physical laws of universe which when tempestuous were often cruelly devastating. High seas combined with nature's fog and night-hidden rocks were uncompromising.

And it followed that these Great Pirates came into mortal battle with one another to see who was going to control the vast sea routes and eventually the world. Their battles took place out of sight of landed humanity. Most of the losers went to the bottom utterly unbeknownst to historians. Those who stayed on the top of the waters and prospered did so because of their comprehensive capability. That is they were the antithesis of specialists. They had high proficiency in dealing with celestial navigation, the storms, the sea, the men, the ship, economics, biology, geography, history, and science. The wider and more long distanced their anticipatory strategy, the more successful they became.

But these hard, powerful, brilliantly resourceful sea masters had to sleep occasionally, and therefore found it necessary to surround themselves with super-loyal, muscular but dull-brained illiterates who could not see nor savvy their masters' stratagems. There was great safety in the mental dullness of these henchmen. The Great Pirates realized that the only people who could possibly contrive to displace them were the truly bright people. For this reason their number-one strategy was secrecy. If the other powerful pirates did not know where you were going, nor when you had gone, nor when you were coming back, they would not know how to waylay you. If anyone knew when you were coming home, "small-timers" could come out in small boats and waylay you in the dark and take you over—just before you got home tiredly after a two-year treasure-harvesting voyage. Thus hijacking and second-rate piracy became a popular activity around the world's shores and harbors. Thus secrecy became the essence of the lives of the successful pirates; ergo, how little is known today of that which I am relating.

Leonardo da Vinci is the outstanding example of the comprehensively anticipatory design scientist. Operating under the patronage of the Duke of Milan he designed the fortified defences and weaponry as well as the tools of peaceful production. Many other great military powers had their comprehensive design scientist-artist inventors; Michelangelo was one of them.

Many persons wonder why we do not have such men today. It is a mistake to think we cannot. What happened at the time of Leonardo and Galileo was that mathematics was so improved by the advent of the zero that not only was much more scientific shipbuilding made possible but also much more reliable navigation. Immediately thereafter truly large-scale venturing on the world's oceans commenced, and the strong sword-leader patrons as admirals put their Leonardos to work, first in designing their new and more powerful world-girdling ships. Next they took their Leonardos to sea with them as their seagoing Merlins to invent ever more powerful tools and strategies on a world-around basis to implement their great campaigns to best all the other great pirates, thereby enabling them to become masters of the world and of all its people and wealth. The required and scientifically designed secrecy of the sea operations thus pulled a curtain that hid the Leonardos from public view, popular ken, and recorded history.

Finally, the sea-dwelling Leonardos became Captains of the ships or even Admirals of Fleets, or Commandants of the Navy yards where they designed and built the fleets, or they became the commandants of the naval war colleges where they designed and developed the comprehensive strategy for running the world for a century to come. This included not only the designing of the network of world-around voyaging and of the ships for each task but also the designing of the industrial establishments and world-around mining operations and naval base-building for production and maintenance of the ships. This Leonardo-type planning inaugurated today's large-scale, world-around industrialization's vast scale of thinking. When the Great Pirates came to building steel steamships and blast furnaces and railroad tracks to handle the logistics, the Leonardos appeared momentarily again in such men as Telford who built the railroads, tunnels, and bridges of England, as well as the first great steamship.

You may say, "Aren't you talking about the British Empire?" I answer, No! The so-called British Empire was a manifest of the world-around misconception of who ran things and a disclosure of the popular ignorance of the Great Pirates' absolute world-controlling through their local-stooge sovereigns and their prime ministers, as only innocuously and locally modified here and there by the separate sovereignties' internal democratic processes. As we soon shall see, the British Isles lying off the coast of Europe constituted in effect a fleet of unsinkable ships and naval bases commanding all the great harbours of Europe. Those islands were the possession of the topmost Pirates. Since the Great Pirates were building, maintaining, supplying their ships on those islands, they also logically made up their crews out of the native islanders who were simply seized or commanded aboard by imperial edict. Seeing these British Islanders aboard the top pirate ships the people around the world mistakenly assumed that the world conquest by the Great Pirates was a conquest by the will, ambition, and organization of the British people. Thus was the G. P.'s grand deception victorious. But the people of those islands never had the ambition to go out and conquer the world. As a people they were manipulated by the top pirates and learned to cheer as they were told of their nation's world prowess.

The topmost Great Pirates' Leonardos discovered—both in their careful, long-distance planning and in their anticipatory inventing—that the grand strategies of sea power made it experimentally clear that a plurality of ships could usually outmaneuver one ship. So the Great Pirates' Leonardos invented navies. Then, of course, they had to control various resource-supplying mines, forests, and lands with which and upon which to build the ships and establish the industries essential to building, supplying, and maintaining their navy's ships.

Then came the grand strategy which said, "divide and conquer." You divide up the other man's ships in battle or you best him when several of his ships are hauled out on the land for repairs. They also had a grand strategy of *anticipatory divide and conquer. Anticipatory divide and conquer* was much more effective than *tardy divide and conquer,* since it enabled those who employed it to surprise the other pirate under conditions unfavorable to the latter. So the great top pirates of the world, realizing that dull people were innocuous and that the only people who could contrive to displace the supreme pirates were the bright ones, set about to apply their grand strategy of *anticipatory divide and conquer* to solve that situation comprehensively.

The Great Pirate came into each of the various lands where he either acquired or sold goods profitably and picked the strongest man there to be his local head man. The Pirate's picked man became the Pirate's general manager of the local realm. If the Great Pirate's local strong man in a given land had not already done so, the Great Pirate told him to proclaim himself king. Despite the local head man's secret subservience to him, the Great Pirate allowed and counted upon his king-stooge to convince his countrymen that he, the local king, was indeed the head man of all men —the god–ordained ruler. To guarantee that sovereign claim the Pirates gave their stooge-kings secret lines of supplies which provided everything required to enforce the sovereign claim. The more massively bejewelled the king's gold crown, and the more visible his court and castle, the less visible was his pirate master.

The Great Pirates said to all their lieutenants around the world, "Any time bright young people show up, I'd like to know about it, because we need bright men." So each time the Pirate came into port the local king-ruler would mention that he had some bright, young men whose capabilities and thinking shone out in the community. The Great Pirate would say to the king, "All right, you summon them and deal with them as follows: As each young man is brought forward you say to him, 'Young man, you are very bright. I'm going to assign you to a great history tutor and in due course if you study well and learn enough I'm going to make you my Royal Historian, but you've got to pass many examinations by both your teacher and myself.'" And when the next bright boy was brought before him the King was to say, "I'm going to make you my Royal Treasurer," and so forth. Then the Pirate said to the king, "You will finally say to all of them: 'But each of you must mind your own business or off go your heads. I'm the only one who minds everybody's business.'"

And this is the way schools began—as the royal tutorial schools. You realize, I hope, that I am not being facetious. That is *it*. This is the beginning of schools and colleges and the beginning of *intellectual specialization*. Of course, it took great wealth to start schools, to have great teachers, and to house, clothe, feed, and cultivate both teachers and students. Only the Great-Pirate-protected robber-barons and the Pirate-protected and secret intelligence-exploited international religious organizations could afford such scholarship investment. And the development of the bright ones into specialists gave the king very great brain power, and made him and his kingdom the most powerful in the land and thus, secretly and greatly, advantaged his patron Pirate in the world competition with the other Great Pirates.

But specialization is in fact only a fancy form of slavery wherein the "expert" is fooled into accepting his slavery by making him feel that in return he is in a socially and culturally preferred, ergo, highly secure, lifelong position. But only the king's son received the kingdom-wide scope of training.

However, the big thinking in general of a spherical Earth and celestial navigation was retained exclusively by the Great Pirates, in contradistinction to a four-cornered, flat world concept, with empire and kingdom circumscribed knowledge, constricted to only that which could be learned through localized preoccupations. Knowledge of the world and its resources was enjoyed exclusively by the Great Pirates, as were also the arts of navigation, shipbuilding and handling, and of grand logistical strategies and of nationally-undetectable, therefore effectively deceptive, international exchange media and trade balancing tricks by which the top pirate, as (in gambler's parlance) "the house," always won.

3 *comprehensively commanded automation*

THEN there came a time, which was World War I, when the most powerful out-pirates challenged the in-pirates with the scientific and technological innovation of an entirely new geometry of thinking. The out-pirates attack went under and above the sea surface and into the invisible realm of electronics and chemical warfaring. Caught off-guard, the in-pirates, in order to save themselves, had to allow their scientists to go to work on their own inscrutable terms. Thus, in saving themselves, the Great Pirates allowed the scientists to plunge their grand, industrial logistics, support strategy into the vast ranges of the electro-magnetic spectrum that were utterly invisible to the pirates.

The pirates until then had ruled the world through their extraordinarily keen senses. They judged things for themselves, and they didn't trust anyone else's eyes. They trusted only that which they could personally smell, hear, touch, or see. But the Great Pirates couldn't see what was going on in the vast ranges of the electro-magnetic reality. Technology was going from wire to wireless, from track to trackless, from pipe to pipeless, and from visible structural muscle to the invisible chemical element strengths of metallic alloys and electro-magnetics.

The Great Pirates came out of that first world war unable to cope knowledgeably with what was going on in the advanced scientific frontiers of industry. The pirates delegated inspection to their "trouble-shooter" experts, but had to content themselves with relayed secone-hand information. This forced them to appraise blindly—ergo, only opinionatedly—whether this or that man really knew what he was talking about, for the G. P.'s couldn't judge for themselves. Thus the Great Pirates were no longer the masters. That was the end. The Great Pirates became extinct. But because the

G. P.'s had always operated secretly, and because they hoped they were not through, they of course did not announce or allow it to be announced that they were extinct. And because the public had never known of them and had been fooled into thinking of their kingly stooges and local politicians as being in reality the head men, society was and is as yet unaware either that the Great Pirates once ran the world or that they are now utterly extinct.

Though the pirates are extinct, all of our international trade balancing and money ratings, as well as all economic accounting, in both the capitalistic and communistic countries, hold strictly to the rules, value systems, terminology, and concepts established by those Great Pirates. Powerful though many successors to the Great Pirates' fragmented dominions may be, no one government, religion, or enterprise now holds the world's physical or metaphysical initiatives.

The metaphysical initiative, too, has gone into competitive confusion between old religions and more recent political or scientific ideologies. These competitors are already so heavily weighted with physical investments and proprietary expediencies as to vitiate any metaphysical initiative. A new, physically uncompromised, metaphysical initiative of unbiased integrity could unify the world. It could and probably will be provided by the utterly impersonal problem solutions of the computers. Only to their superhuman range of calculative capabilities can and may all political, scientific, and religious leaders face-savingly acquiesce.

Abraham Lincoln's concept of "right triumphing over might" was realized when Einstein as metaphysical intellect wrote the equation of physical universe $E = Mc^2$ and thus comprehended it. Thus the metaphysical took the measure of, and mastered, the physical. That relationship seems by experience to be irreversible. Nothing in our experience suggests that energy could comprehend and write the equation of intellect. That equation is operating inexorably, and the metaphysical is now manifesting its ability to reign over the physical.

This is the essence of human evolution upon Spaceship Earth. If the present planting of humanity upon Spaceship Earth cannot comprehend this inexorable process and discipline itself to serve exclusively that function of metaphysical mastering of the physical it will be discontinued, and its potential mission in universe will be carried on by the metaphysically endowed capabilities of other beings on other spaceship planets of universe.

The Great Pirates did run the world. They

were the first and last to do so. They were world men, and they ran the world with ruthless and brilliant pragmatism based on the mis-seemingly "fundamental" information of their scientifically specialized servants. First came their Royal Society scientific servants, with their "Great" Second Law of thermodynamics, whose "entropy" showed that every energy machine kept losing energy and eventually "ran down." In their pre-speed-of-light-measurement misconceptioning of an omni-simultaneous—*"instant universe"*—that universe, as an energy machine was thought, also to be "running down." And thus the energy wealth and life support were erroneously thought to be in continuous depletion—orginating the misconception of "spending."

Next came Thomas Malthus, professor of political economics of the Great Pirate's East India Company, who said that man was multiplying himself at a geometrical rate and that food was multiplying only at an arithmetical rate. And lastly, thirty-five years later, came the G. P.'s biological specialist servant, Charles Darwin, who, explaining his theory of animate evolution, said that survival was only for the fittest.

Quite clearly to the Great Pirates it was a scientific fact that not only was there not enough to go around but apparently not enough to go around for even 1 per cent of humanity to live at a satisfactorily-sustaining standard of living. And because of entropy the inadequacy would always increase. Hence, said the G. P.'s, survival was obviously a cruel and almost hopeless battle. They ran the world on the basis that these Malthusian-Darwinian entropy concepts were absolute scientific laws, for that was what their scientifically respected, intellectual slave specialists had told them.

Then we have the great pragmatic ideologist Marx running into that entropic-Malthusian-Darwinian information and saying, "Well, the workers who produce things are the fittest because they are the only ones who know how to physically produce and therefore they ought to be the ones to survive." That was the beginning of the great "class warfare." All of the ideologies range somewhere between the Great Pirates and the Marxists. But all of them assume that there is not enough to go around. And that's been the rationalized working hypothesis of all the great sovereign claims to great areas of the Earth. Because of their respective exclusivities, all the class warfare ideologies have become extinct. Capitalism and socialism are mutually extinct. Why? Because science now finds there can be ample for all, but only if the sovereign fences are completely removed. The basic you-or-me-not-enough-for-both—ergo, someone-must-die– tenets of the class warfaring are extinct.

Now let us examine more closely what we know scientifically about extinction. At the annual Congress of the American Association for the Advancement of Science, as held approximately ten years ago in Philadelphia, two papers were presented in widely-separated parts of the Congress. One was presented in anthropology and the other in biology, and though the two author-scientists knew nothing of each other's efforts they were closely related. The one in anthropology examined the case histories of all the known human tribes that had become extinct. The biological paper investigated the case histories of all the known biological species that had become extinct. Both scientists sought for a common cause of extinction. Both of them found a cause, and when the two papers were accidentally brought together it was discovered that the researchers had found the same causes. Extinction in both cases was the consequence of over-specialization. How does that come about?

We can develop faster and faster running horses as specialists. To do so we inbreed by mating two fast-running horses. By concentrating certain genes the probability of their dominance is increased. But in doing so we breed out or sacrifice general adaptability. Inbreeding and specialization always do away with general adaptability.

There's a major pattern of energy in universe wherein the very large events, earthquakes, and so forth, occur in any one area of universe very much less frequently than do the small energy events. Around the Earth insects occur more often than do earthquakes. In the patterning of total evolutionary events, there comes a time, once in a while, amongst the myriad of low energy events, when a large energy event transpires and is so disturbing that with their general adaptability lost, the ultra-specialized creatures perish. I will give you a typical history—that of a type of bird which lived on a special variety of micro-marine life. Flying around, these birds gradually discovered that there were certain places in which that particular marine life tended to pocket—in the marshes along certain ocean shores of certain lands. So, instead of flying aimlessly for chance finding of that marine life they went to where it was concentrated in bayside marshes. After a while, the water began to recede in the marshes, because the Earth's polar ice cap was beginning to increase. Only the birds with very long beaks could reach deeply enough in the marsh holes to get at the marine life. The unfed, short-billed birds died off. This left only the long-beakers. When the birds' inborn drive to reproduce occurred there were only other long-beakers surviving with whom to breed. This concentrated their long-beak genes. So, with continually receding waters and generation to generation inbreeding, longer and longer beaked birds were produced. The waters kept receding, and the beaks of successive generations of the birds grew bigger and bigger. The long-beakers seemed to be prospering when all at once there was a great fire in the marshes. It was discovered that because their beaks had become so heavy these birds could no longer fly. They could not escape the flames by flying out of the marsh. Waddling on their legs they were too slow to escape, and so they perished. This is typical of the way in which extinction occurs—through over-specialization.

When, as we have seen, the Great Pirates let their scientists have free rein in World War I the Pirates themselves became so preoccupied with enormous wealth harvesting that they not only lost track of what the scientists were doing within the vast invisible world but they inadvertently abandoned their own comprehensivity and they, too, became severe specialists as industrial production money makers, and thus they compounded their own acceleration to extinction in the world-paralyzing economic crash of 1929. But society, as we have seen, never knew that the Great Pirates had been running the world. Nor did society realize in 1929 that the Great Pirates had become extinct. However, world society was fully and painfully aware of the economic paralysis. Society consisted then, as now, almost entirely of specialized slaves in education, management, science, office routines, craft, farming, pick-and-shovel labour, and their families. Our world society now has none of the comprehensive and realistic world knowledge that the Great Pirates had.

Because world societies thought mistakenly of their local politicians, who were only the stooges of the Great Pirates, as being realistically the head men, society went to them to get the industrial and economic machinery going again. Because industry is inherently world-coordinate these world economic depression events of the 1920's and 1930's meant that each of the local head politicians of a number of countries were asked separately to make the world work. On this basis the world-around inventory of resources was no longer integratable. Each of the political leaders' mandates were given from different ideological groups, and their differing viewpoints and resource difficulties led inevitably to World War II.

The politicians, having an automatic bias, were committed to defend and advantage only their own side. Each assumed the validity of the Malthusian-Darwin-you-or-me-to-the-death struggle. Because of the working concept that there was not enough to go around, the most aggressive political leaders exercised their political leadership by heading their countries into war to overcome the rest of the world, thus to dispose of the unsupportable excess population through decimation and starvation—the age-old, lethal formula of ignorant men. Thus we had all our world society specializing, whether under fascism, communism, or capitalism. All the great ideological groups assumed Armageddon.

Getting ready for the assumed inexorable Armageddon, each applied science and all of the great scientific specialization capabilities only toward weaponry, thus developing the ability to destroy themselves totally with no comprehensively organized oppositional thinking capability and initiative powerful enough to co-ordinate and prevent it. Thus by 1946, we were on the swift way to extinction despite the inauguration of the United Nations, to which none of the exclusive sovereign prerogatives were surrendered. Suddenly, all unrecognized as such by society, the evolutionary antibody to the extinction of humanity through specialization appeared in the form of the computer and its comprehensively commanded automation which made man obsolete as a physical production and control specialist—and just in time.

The computer as superspecialist can persevere, day and night, day after day, in picking out the pink from the blue at superhumanly sustainable speeds. The computer can also operate in degrees of cold or heat at which man would perish. Man is going to be displaced altogether as a specialist by the computer. Man himself is being forced to reestablish, employ, and enjoy his innate "comprehensivity." Coping with the totality of Spaceship Earth and universe is ahead for all of us. Evolution is apparently intent that man fulfill a much greater destiny than that of being a simple muscle and reflex machine—a slave automaton—*automation* displaces the *automatons.*

Evolution consists of many great revolutionary events taking place quite independently of man's consciously attempting to bring them about. Man is very vain; he likes to feel that he is responsible for all the favorable things that happen, and he is innocent of all the unfavorable happenings. But all the larger evolutionary patternings seeming favorable or unfavorable to man's conditioned reflexing are transpiring transcendentally to any of man's conscious planning or contriving.

To disclose to you your own vanity of reflexing, I remind you quickly that none of you is consciously routing the fish and potato you ate for lunch into this and that specific gland to make hair, skin, or anything like that. None of you are aware of how you came to grow from 7 pounds to 70 pounds and then to 170 pounds, and so forth. All of this is automated, and always has been. There is a great deal that is automated regarding our total salvation on Earth, and I would like to get in that frame of mind right now in order to be useful in the short time we have.

Let us now exercise our intellectual faculties as best we can to apprehend the evolutionary patternings transcending our spontaneous cognitions and recognitions. We may first note an evolutionary trend that countered all of the educational systems and the deliberately increased professional specialization of scientists. This contradiction occurred at the beginning of World War II, when extraordinary new scientific instruments had been developed and the biologists and chemists and physicists were meeting in Washington, D. C., on special war missions. Those scientists began to realize that whereas a biologist used to think he was dealing only in cells and that a chemist was dealing only in molecules and the physicist was dealing only in atoms, they now found their new powerful instrumentation and contiguous operations overlapping. Each specialist suddenly realized that he was concerned alike with atoms, molecules, and cells. They found there was no real dividing line between their professional interests. They hadn't meant to do this, but their professional fields were being integrated —inadvertently, on their part, but apparently purposefully—by inexorable evolution. So, as of World War II, the scientists began to invert new professional designations: the bio-chemist, the bio-physicist, and so forth. They were forced to. Despite their deliberate attempts only to specialize, they were being merged into ever more inclusive fields of consideration. Thus was deliberately specializing man led back unwittingly once more to reemploy his innately comprehensive capabilities.

I find it very important in disembarrassing ourselves of our vanity, short-sightedness, biases, and ignorance in general, in respect to universal evolution, to think in the following manner. I've often heard people say, "I wonder what it would be like to be on board a spaceship," and the answer is very simple. What *does* it *feel* like? That's all we have ever experienced. We are all astronauts.

I know you are paying attention, but I'm sure you don't immediately agree and say, "Yes, that's right, I am an astronaut." I'm sure that you don't really sense yourself to be aboard a fantastically real spaceship—our spherical

Spaceship Earth. Of our little sphere you have seen only small portions. However, you have viewed more than did pre-twentieth-century man, for in his entire lifetime he saw only one-millionth of the Earth's surface. You've seen a lot more. If you are a veteran world airlines pilot you may have seen one one-hundredth of Earth's surface. But even that is sum totally not enough to see and feel Earth to be a sphere—unless, unbeknownst to me, one of you happens to be a Cape Kennedy capsuler.

4 *spaceship earth*

OUR little Spaceship Earth is only eight thousand miles in diameter, which is almost a negligible dimension in the great vastness of space. Our nearest star—our energy-supplying mother-ship, the Sun—is ninety-two million miles away, and the next nearest star is one hundred thousand times further away. It takes two and one-half years for light to get to us from the next nearest energy supply ship star. That is the kind of space-distanced pattern we are flying. Our little Spaceship Earth is right now travelling at sixty thousand miles an hour around the sun and is also spinning axially, which, at the latitude of Washington, D. C., adds approximately one thousand miles per hour to our motion. Each minute we both spin at one hundred miles and zip in orbit at one thousand miles. That is a whole lot of spin and zip. When we launch our rocketed space capsules at fifteen thousand miles an hour, that additional acceleration speed we give the rocket to attain its own orbit around our speeding Spaceship Earth is only one-fourth greater than the speed of our big planetary spaceship.

Spaceship Earth was so extraordinarily well invented and designed that to our knowledge humans have been on board it for two million years not even knowing that they were on board a ship. And our spaceship is so superbly designed as to be able to keep life regenerating on board despite the phenomenon, entropy, by which all local physical systems lose energy. So we have to obtain our biological life-regenerating energy from another spaceship–the sun.

Our sun is flying in company with us, within the vast reaches of the Galactic system, at just the right distance to give us enough radiation to keep us alive, yet not close enough to burn us up. And the whole scheme of Spaceship Earth and its live passengers is so superbly designed that the Van Allen belts, which we didn't even know we had until yesterday, filter the sun and other star radiation which as it impinges upon our spherical ramparts is so concentrated that if we went nakedly outside the Van Allen belts it would kill us. Our Spaceship Earth's designed infusion of that radiant energy of the stars is processed in such a way that you and I can carry on safely. You and I can go out and take a sunbath, but are unable to take in enough energy through our skins to keep alive. So part of the invention of the Spaceship Earth and its biological life-sustaining is that the vegetation on the land and the algae in the sea, employing photosynthesis, are designed to impound the life-regenerating energy for us to adequate amount.

But we can't eat all the vegetation. As a matter of fact, we can eat very little of it. We can't eat the bark nor wood of the trees nor the grasses. But insects can eat these, and there are many other animals and creatures that can. We get the energy relayed to us by taking the milk and meat from the animals. The animals can eat the vegetation, and there are a few of the fruits and tender vegetation petals and seeds that we can eat. We have learned to cultivate more of those botanical edibles by genetical inbreeding.

That we are endowed with such intuitive and intellectual capabilities as that of discovering the genes and the R.N.A. and D.N.A. and other fundamental principles governing the fundamental design controls of life systems as well as of nuclear energy and chemical structuring is part of the extraordinary design of the Spaceship Earth, its equipment, passengers, and internal support systems. It is therefore paradoxical but strategically explicable, as we shall see, that up to now we have been mis-using, abusing, and polluting this extraordinary chemical energy-interchanging system for successfully regenerating all life aboard our planetary spaceship.

One of the interesting things to me about our spaceship is that it is a mechanical vehicle, just as is an automobile. If you own an automobile, you realize that you must put oil and gas into it, and you must put water in the radiator and take care of the car as a whole. You begin to develop quite a little thermodynamic sense. You know that you're either going to have to keep the machine in good order or it's going to be in trouble and fail to function. We have not been seeing our Spaceship Earth as an integrally-designed machine which to be persistently successful must be comprehended and serviced in total.

Now there is one outstandingly important fact regarding Spaceship Earth, and that is that no instruction book came with it. I think it's very significant that there is no instruction book for successfully operating our ship. In view of the infinite attention to all other details displayed by our ship, it must be taken as deliberate and purposeful that an instruction book was omitted. Lack of instruction has forced us to find that there are two kinds of berries—red berries that will kill us and red berries that will nourish us. And we had to find out ways of telling which-was-which red berry before we ate it or otherwise we would die. So we were forced, because of a lack of an instruction book, to use our intellect, which is our supreme faculty, to devise scientific experimental procedures and to interpret effectively the significance of the experimental findings. Thus, because the instruction manual was missing we are learning how we safely can anticipate the consequences of an increasing number of alternative ways of extending our satisfactory survival and growth—both physical and metaphysical.

Quite clearly, all of life as designed and born is utterly helpless at the moment of birth. The human child stays helpless longer than does the young of any other species. Apparently it is part of the invention "man" that he is meant to be utterly helpless through certain anthropological phases and that, when he begins to be able to get on a little better, he is meant to discover some of the physical leverage-multiplying principles inherent in universe as well as the many nonobvious resources around him which will further compoundingly multiply his knowledge-regenerating and life-fostering advantages.

I would say that designed into this Spaceship Earth's total wealth was a big safety factor which allowed man to be very ignorant for a long time until he had amassed enough experiences from which to extract progressively the system of generalized principles governing the increases of energy managing advantages over environment. The designed omission of the instruction book on how to operate and maintain Spaceship Earth and its complex life-supporting and regenerating systems has forced man to discover retrospectively just what his most important forward capabilities are. His intellect had to discover itself. Intellect in turn had to compound the facts of his experience. Comprehensive reviews of the compounded facts of experiences by intellect brought forth awareness of the generalized principles underlying all special and only superficially-sensed experiences. Objective employment of those generalized principles in rearranging the physical resources of environment seems to be leading to humanity's eventually total success and readiness to cope with far vaster problems of universe.

To comprehend this total scheme we note that long ago a man went through the woods, as you may have done, and I certainly have, trying to find the shortest way through the woods in a given direction. He found trees fallen across his path. He climbed over those crisscrossed trees and suddenly found himself poised on a tree that was slowly teetering. It happened to be lying across another great tree, and the other end of the tree on which he found himself teetering lay under a third great fallen tree. As he teetered he saw the third big tree lifting. It seemed impossible to him. He went over and tried using his own muscles to lift that great tree. He couldn't budge it. Then he climbed back atop the first smaller tree, purposefully teetering it, and surely enough it again elevated the larger tree. I'm certain that the first man who found such a tree thought that it was a magic tree, and may have dragged it home and erected it as man's first totem. It was probably a long time before he learned that any stout tree would do, and thus extracted the concept of the generalized principle of leverage out of all his earlier successive special-case experiences with such accidental discoveries. Only as he learned to generalize fundamental principles of physical universe did man learn to use his intellect effectively.

Once man comprehended that any tree would serve as a lever his intellectual advantages accelerated. Man freed of special-case superstition by intellect has had his survival potentials multiplied millions fold. By virtue of the leverage principles in gears, pulleys, transistors, and so forth, it is literally possible to do more with less in a multitude of physio-chemical ways. Possibly it was this intellectual augmentation of humanity's survival and success through the metaphysical perception of generalized principles which may be objectively employed that Christ was trying to teach in the obscurely told story of the loaves and the fishes.

5 general systems theory

HOW may we use our intellectual capability to higher advantage? Our muscle is very meager as compared to the muscles of many animals. Our integral muscles are as nothing compared to the power of a tornado or the atom bomb which society contrived–in fear–out of the intellect's fearless discoveries of generalized principles governing the fundamental energy behaviors of physical universe.

In organizing our grand strategy we must first discover where we are now; that is, what our present navigational position in the universal scheme of evolution is. To begin our position-fixing aboard our Spaceship Earth we must first acknowledge that the abundance of immediately consumable, obviously desirable or utterly essential resources have been sufficient until now to allow us to carry on despite our ignorance. Being eventually exhaustible and spoilable, they have been adequate only up to this critical moment. This cushion-for-error of humanity's survival and growth up to now was apparently provided just as a bird inside of the egg is provided with liquid nutriment to develop it to a certain point. But then by design the nutriment is exhausted at just the time when the chick is large enough to be able to locomote on its own legs. And so as the chick pecks at the shell seeking more nutriment it inadvertently breaks open the shell. Stepping forth from its initial sanctuary, the young bird must now forage on its own legs and wings to discover the next phase of its regenerative sustenance.

My own picture of humanity today finds us just about to step out from amongst the pieces of our just one-second-ago broken eggshell. Our innocent, trial-and-error-sustaining nutriment is exhausted. We are faced with an entirely new relationship to the universe. We are going to have to spread our wings of intellect and fly or perish; that is, we must dare immediately to fly by the generalized principles governing universe and not by the ground rules of yesterday's superstitious and erroneously conditioned reflexes. And as we attempt competent thinking we immediately begin to reemploy our innate drive for comprehensive understanding.

The architects and planners, particularly the planners, though rated as specialists, have a little wider focus than do the other professions. Also as human beings they often battle the narrow views of specialists–in particular, their patrons–the politicians, and the financial and other legal, but no longer comprehensively effective, heirs to the great pirates'–now only ghostly–prerogatives. At least the planners are allowed to look at *all* of Philadelphia, and not just to peek through a hole at one house or through one door at one room in that house. So I think it's appropriate that we assume the role of planners and begin to do the largest scale comprehensive thinking of which we are capable.

We begin by eschewing the role of specialists who deal only in parts. Becoming deliberately expansive instead of contractive, we ask, "*How* do we think in terms of *wholes*?" If it is true that the bigger the thinking becomes the more lastingly effective it is, we must ask, "How big can we think?"

One of the modern tools of high intellectual advantage is the development of what is called general systems theory. Employing it we begin to think of the largest and most comprehensive systems, and try to do so scientifically. We start by inventorying all the important, known variables that are operative in the problem. But if we don't really know how big "big" is, we may not start big enough, and are thus likely to leave unknown, but critical, variables outside the system which will continue to plague us. Interaction of the unknown variables inside and outside the arbitrarily chosen limits of the system are probably going to generate misleading or outrightly wrong answers. If we are to be effective, we are going to have to think in both the biggest and most minutely-incisive ways permitted by intellect and by the information thus far won through experience.

Can we think of, and state adequately and incisively, what we mean by universe? For universe is, inferentially, the biggest system. If we could start with universe, we would automatically avoid leaving out any strategically critical variables. We find no record as yet of man having successfully defined the universe—scientifically and comprehensively—to include the nonsimultaneous and only partially overlapping, micro-macro, always and everywhere transforming, physical and metaphysical, omni-complementary but nonidentical events.

Man has failed thus far, as a specialist, to define the microcosmic limits of divisibility of the nucleus of the atom, but, epochally, as accomplished by Einstein, has been able to define successfully the physical universe but not the metaphysical universe; nor has he, as yet, defined total universe itself as combining both the physical and metaphysical. The scientist was able to define physical universe by virtue of the experimentally-verified discovery that energy can neither be created nor lost and, therefore, that energy is conserved and is therefore finite. That means it is equatable.

Einstein successfully equated the physical universe as $E = Mc^2$. His definition was only a hypothetical venture until fission proved it to be true. The physical universe of associative and disassociative energy was found to be a closed, but nonsimultaneously occurring, system—its separately occurring events being mathematically measurable; i.e., weighable and equatable. But the finite physical universe did not include the metaphysical weightless experiences of universe. All the unweighables, such as any and all our thoughts and all the abstract mathematics, are weightless. The metaphysical aspects of universe have been thought by the physical scientists to defy "closed system's" analysis. I have found, however, as we shall soon witness, that total universe including both its physical and metaphysical behaviors and aspects are scientifically definable.

Einstein and others have spoken exclusively about the physical department of universe in words which may be integrated and digested as *the aggregate of nonsimultaneous and only partially overlapping, nonidentical, but always complementary, omni-transforming, and weighable energy events.* Eddington defines science as "the earnest attempt to set in order the facts of experience." Einstein and many other first-rank scientists noted that science is concerned exclusively with "facts of experience."

Holding to the scientists' experiences as all important, I define universe, including both the physical and metaphysical, as follows: *The universe is the aggregate of all of humanity's consciously-apprehended and communicated experience with the nonsimultaneous, nonidentical, and only partially overlapping, always complementary, weighable and unweighable, ever omni-transforming, event sequences.*

Each experience begins and ends—ergo, is finite. Because our apprehending is packaged, both physically and metaphysically, into time increments of alternate awakeness and asleepness as well as into separate finite conceptions such as the discrete energy quanta and the atomic nucleus components of the fundamental physical discontinuity, all experiences are finite. Physical experiments have found no solids, no continuous surfaces or lines—only discontinuous constellations of individual events. An aggregate of finites is finite. *Therefore, universe as experientially defined, including both the physical and metaphysical, is finite.*

It is therefore possible to initiate our general systems formulation at the all inclusive level of universe whereby no strategic variables will be omitted. There is an operational grand strategy of General Systems Analysis that proceeds from here. It is played somewhat like the game of "Twenty Questions," but G. S. A. is more efficient—that is, is more economical—in reaching its answers. It is the same procedural strategy that is used by the computer to weed out all the wrong answers until only the right answer is left.

Having adequately defined the whole system we may proceed to subdivide progressively. This is accomplished through progressive division into two parts—one of which, by definition, could not contain the answer—and discarding of the sterile part. Each progressively-retained live part is called a "bit" because of its being produced by the progressive binary "yes" or "no" bi-section of the previously residual live part. The magnitude of such weeding operations is determined by the number of successive bits necessary to isolate the answer.

How many "bi-secting bits" does it take to get rid of all the irrelevancies and leave in lucid isolation that specific information you are seeking? We find that the first subdividing of the concept of universe–bit one–is into what we call a *system. A system subdivides universe into all the universe outside the system* (*macrocosm*) *and all the rest of the universe which is inside the system* (*microcosm*) *with the exception of the minor fraction of universe which constitutes the system itself.* The system divides universe not only into macrocosm and microcosm but also coincidentally into typical conceptual and nonconceptual aspects of universe–that is, an overlappingly-associable consideration, on the one hand, and, on the other hand, all the nonassociable, nonoverlappingly-considerable, nonsimultaneously-transforming events of nonsynchronizable disparate wave frequency rate ranges.

A thought is a system, and is inherently conceptual–though often only dimly and confusedly conceptual at the moment of first awareness of the as yet only vaguely describable thinking activity. Because total universe is nonsimultaneous it is not conceptual. Conceptuality is produced by isolation, such as in the instance of one single, static picture held out from a moving-picture film's continuity, or scenario. Universe is an evolutionary-process scenario without beginning or end, because the shown part is continually transformed chemically into fresh film and reexposed to the ever self-reorganizing process of latest thought realizations which must continually introduce new significance into the freshly written description of the ever-transforming events before splicing the film in again for its next projection phase.

Heisenberg's principle of "indeterminism" which recognized the experimental discovery that the act of measuring always alters that which was being measured turns experience into a continuous and never-repeatable evolutionary scenario. One picture of the scenario about the caterpillar phase does not communicate its transformation into the butterfly phase, etc. The question, "I wonder what is outside the outside-of-universe?" is a request for a single picture description of a scenario of transformations and is an inherently invalid question. It is the same as looking at a dictionary and saying, "Which word is the dictionary?" It is a meaningless question.

It is characteristic of "all" thinking—of all system's conceptioning—that all the lines of thought interrelationships must return cyclically upon themselves in a plurality of directions, as do various great circles around spheres. Thus may we interrelatedly comprehend the constellation—or system—of experiences under consideration. Thus may we comprehend how the special-case economy demonstrated by the particular system considered also discloses the generalized law of energy conservation of physical universe.

To hit a duck in flight a hunter does not fire his gun at the bird where the gunner sees him but ahead of the bird, so that the bird and the bullet will meet each other at a point not in line between the gunner and the bird at time of firing. Gravity and wind also pull the bullet in two different directions which altogether impart a mild corkscrew trajectory to the bullet. Two airplanes in nighttime dogfights of World War II firing at each other with tracer bullets and photographed by a third plane show clearly the corkscrew trajectories as one hits the other. Einstein and Reiman, the Hindu mathematician, gave the name *geodesic lines* to these curvilinear and *most economical lines of interrelationship between two independently moving "events"*—the events in this case being the two airplanes.

A great circle is a line formed on a sphere's surface by a plane going through the sphere's centre. Lesser circles are formed on the surfaces of spheres by planes cutting through spheres but not passing through the sphere's centre. When a lesser circle is superimposed on a great circle it cuts across the latter at two points, *A* and *B*. It is a shorter distance between *A* and *B* on the great circle's shortest arc than it is on the lesser circle's shortest arc. Great circles are geodesic lines because they provide the most economical (energy, effort) distances between any two points on a spherical system's surface; therefore, nature, which always employs only the most economical realizations, must use those great circles which, unlike spiral lines, always return upon themselves in the most economical manner. All the system's paths must be topologically and circularly interrelated for conceptually definitive, locally transformable, polyhedronal understanding to be attained in our spontaneous—ergo, most economical— geodesicly structured thoughts.

Thinking itself consists of self-disciplined dismissal of both the macrocosmic and microcosmic irrelevancies which leaves only the lucidly-relevant considerations. The macrocosmic irrelevancies are all the events too large and too infrequent to be synchronizably tuneable in any possible way with our consideration (a beautiful word meaning putting stars to-

UNDERLYING ORDER IN RANDOMNESS

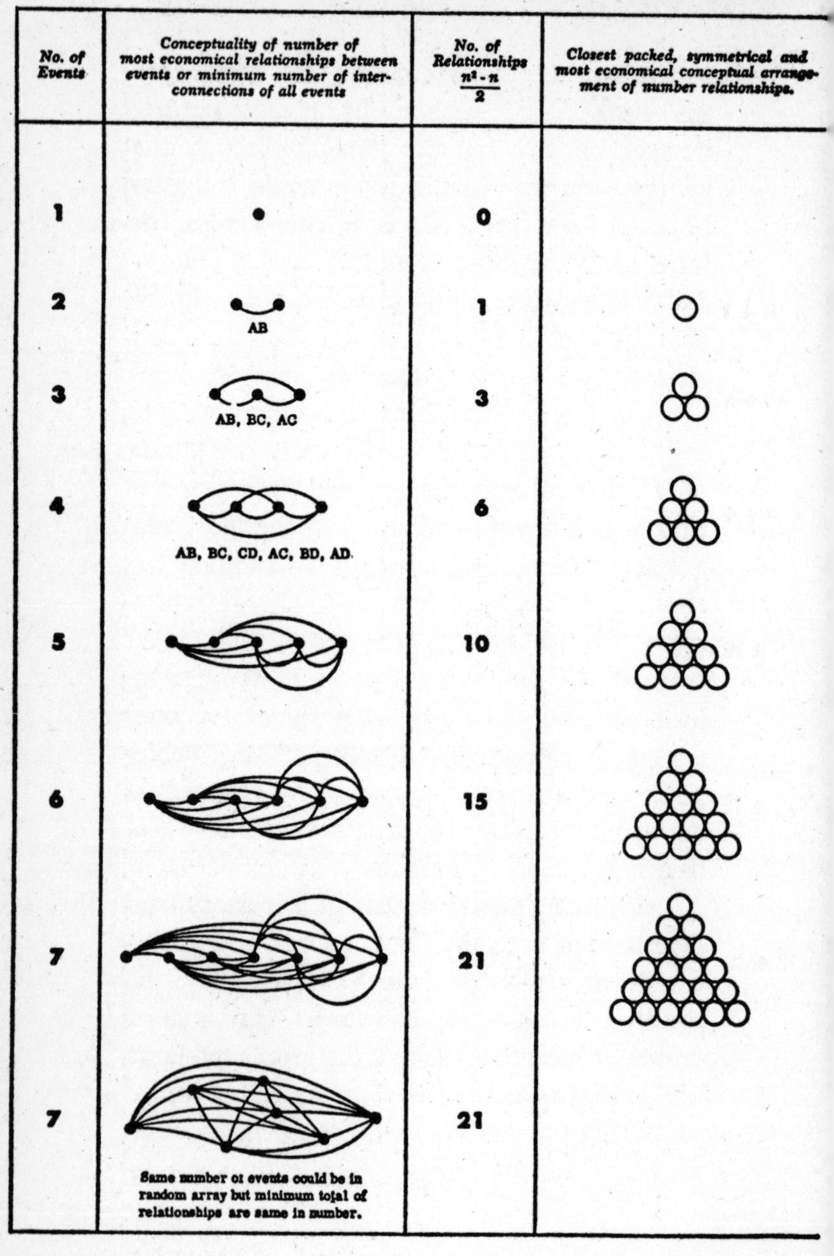

No. of Events	Conceptuality of number of most economical relationships between events or minimum number of interconnections of all events	No. of Relationships $\frac{n^2-n}{2}$	Closest packed, symmetrical and most economical conceptual arrangement of number relationships.
1		0	
2	AB	1	
3	AB, BC, AC	3	
4	AB, BC, CD, AC, BD, AD	6	
5		10	
6		15	
7		21	
7	Same number of events could be in random array but minimum total of relationships are same in number.	21	

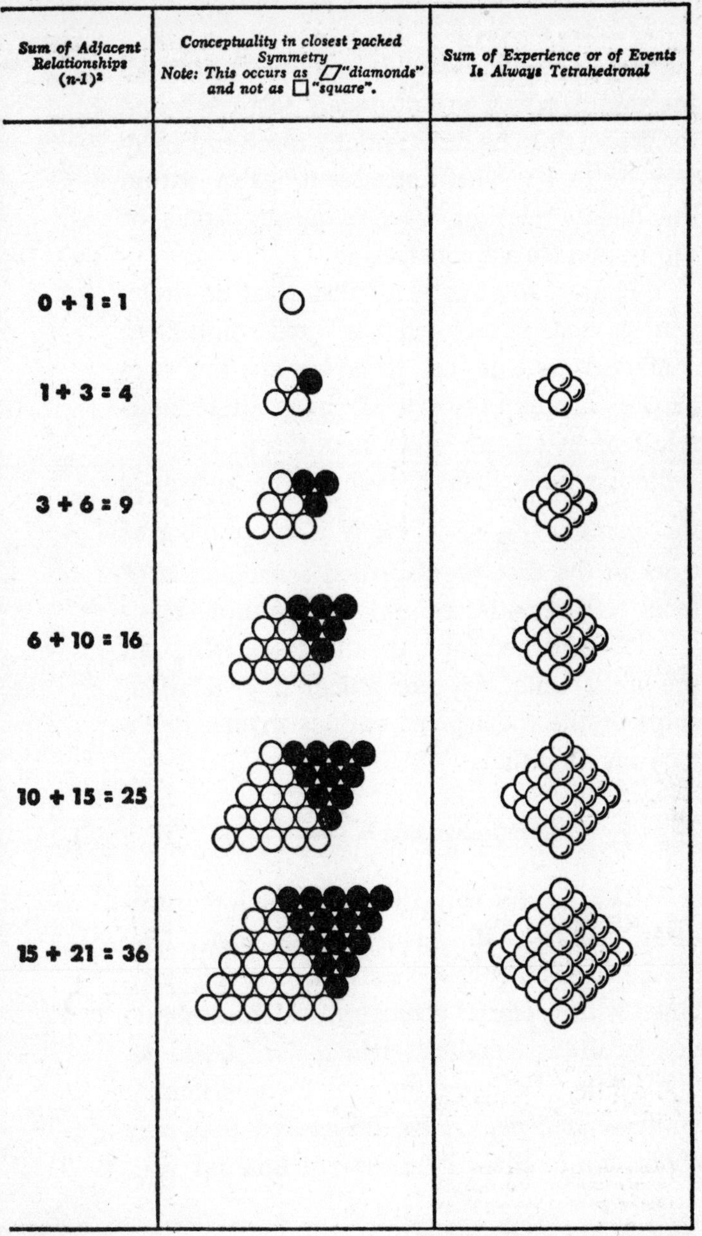
Sum of Adjacent Relationships $(n-1)^2$
Conceptuality in closest packed Symmetry
Note: This occurs as "diamonds" and not as "square".
Sum of Experience or of Events Is Always Tetrahedronal
0 + 1 = 1
1 + 3 = 4
3 + 6 = 9
6 + 10 = 16
10 + 15 = 25
15 + 21 = 36

gether). The microcosmic irrelevancies are all the events which are obviously too small and too frequent to be differentially resolved in any way or to be synchronizably-tuneable within the lucidly-relevant wave-frequency limits of the system we are considering.

How many stages of dismissal of irrelevancies does it take—that is, proceeding from "universe" as I defined it, how many bits does it take—lucidly to isolate all the geodesic inter-relations of all the "star" identities in the constellation under consideration? The answer is the formula $\frac{N^2 - N}{2}$ where N is the number of stars in the thought-discerned constellation of focal point entities comprising the problem.

"Comprehension" means identifying all the most uniquely economical inter-relationships of the focal point entities involved. We may say then that:

$$\text{Comprehension} = \frac{N^2 - N}{2}$$

This is the way in which thought processes operate with mathematical logic. The mathematics involved consist of topology, combined with vectorial geometry, which combination I call "synergetics"—which word I will define while clarifying its use. By questioning many audiences, I have discovered that only about one in three hundred are familiar with synergy. The word is obviously not a popular word. Synergy is the only word in our language that means *behavior of whole systems unpredicted by the separately observed behaviors of any of the system's separate parts or any subassembly of the system's parts.* There is nothing in the chemistry of a toenail that predicts the existence of a human being.

I once asked an audience of the National Honors Society in chemistry, "How many of you are familiar with the word, synergy?" and all hands went up. Synergy is the essence of chemistry. The tensile strength of chrome-nickel, steel, which is approximately 350,000 pounds per square inch, is 100,000 P.S.I. greater than the sum of the tensile strengths of each of all its alloyed together, component, metallic elements. Here is a "chain" that is 50 per cent stronger than the sum of the strengths of all its links. We think popularly only in the terms of a chain being no stronger than its weakest link, which concept fails to consider, for instance, the case of an endlessly interlinked chain of atomically self-renewing links of omni-equal strength or of an omni-directionally interlinked chain matrix of ever renewed atomic links in which one broken link would be, only momentarily, a local cavern within the whole mass having no weakening effect on the whole, for every link within the matrix is a high frequency, recurring, break-and-make restructuring of the system.

Since synergy is the only word in our language meaning behavior of wholes unpredicted by behavior of their parts, it is clear that society does not think there are behaviors of whole systems unpredicted by their separate parts. This means that society's formally-accredited thoughts and ways of accrediting others are grossly inadequate in comprehending the nonconceptual qualities of the scenario "universal evolution."

There is nothing about an electron alone that forecasts the proton, nor is there anything about the Earth or the Moon that forecasts the co-existence of the Sun. The solar system is synergetic—unpredicted by its separate parts. But the interplay of Sun as supply ship of Earth and the Moon's gravitationally produced tidal pulsations on Earth all interact to produce the biosphere's chemical conditions which permit but do not cause the regeneration of life on Spaceship Earth. This is all synergetic. There is nothing about the gases given off respiratorily by Earth's green vegetation that predicts that those gases will be essential to the life support of all mammals aboard Spaceship Earth, and nothing about the mammals that predicts that the gases which they give off respiratorily are essential to the support of the vegetation aboard our Spaceship Earth. Universe is synergetic. Life is synergetic.

Summarizing synergetically I may conclude that since my experimental interrogation of more than one hundred audiences all around the world has shown that less than one in three hundred university students has ever heard of the word synergy, and since it is the only word that has that meaning it is obvious that the world has not thought there are any behaviors of whole systems unpredictable by their parts. This is partially the consequence of over-specialization and of leaving the business of the whole to the old pirates to be visibly conducted by their stooges, the feudal kings or local politicians.

There is a corollary of synergy which says that the known behavior of the whole and the known behavior of a minimum of known parts often makes possible the discovery of the values of the remaining parts as does the known sum of the angles of a triangle plus the known behavior of three of its six parts make possible evaluating the others. Topology provides the synergetic means of ascertaining the values of any system of experiences.

Topology is the science of fundamental pattern and structural relationships of event constellations. It was discovered and developed by the mathematician Euler. He discovered that all patterns can be reduced to three prime conceptual characteristics: to *lines; points* where two lines cross or the same line crosses itself; and *areas*, bound by lines. He found that there is a constant relative abundance of these three fundamentally unique and no further reducible aspects of all patterning

$$P + A = L + 2$$

This reads: the number of points plus the number of areas always equals the number of lines plus the number constant two. There are times when one area happens to coincide with others. When the faces of polyhedra coincide illusionarily the congruently hidden faces must be accounted arithmetically in formula.

WE will now tackle our present world problems with the family of powerful thought tools: *topology, geodesics, synergetics, general systems theory,* and the computer's *operational "bitting."* To insure our inclusion of all the variables with which we must eventually deal, we will always start synergetically with *universe* —now that universe is defined and thus provides us with the *master containment.* We will then state our unique problem and divest ourselves progressively and definitively of all the micro-macro irrelevancies. Are humans necessary? Are the experiential clues that human intellect has an integral function in regenerative universe as has gravity? How can Earthians fulfill their function and thus avoid extinction as unfit?

To start with, we will now progressively subdivide universe and isolate the thinkable concept by bits through progressively dismissing residual irrelevancies. Our first isolated bit is the *system,* which at maximum is the starry macrocosmic and at minimum the atomic nucleus; the second bit reduces the macrocosmic limit to that of the *galactic nebula;* the third bit separates out *cosmic radiation, gravity* and the *solar system;* and the fourth bit isolates the *cosmic radiation, gravity, sun,* its *energized, life-bearing Spaceship Earth,* together WITH the *Earth's Moon* as the most prominent components of the life regeneration on Spaceship Earth.

I would like to inventory rapidly the system variables which I find to be by far the most powerful in the consideration of our present life-regenerating evolution aboard our spaceship as it is continually refueled radiationally by the Sun and other cosmic radiation. Thus we may, by due process, suddenly and excitingly discover why we are here alive in universe and identify ourselves as presently operating here, aboard our spaceship, and situated aboard its spherical deck at, for example, Washington, D. C., on the North American continent, thinking effectively regarding the relevant contemporary and local experiences germane to the solution of humanity's successful and happy survival aboard our planet. We may thus discover not only what needs to be done in a fundamental way but also we may discover how it may be accomplished by our own directly-seized initiative, undertaken and sustained without any further authority than that of our function in universe, where the most ideal is the most realistically practical. Thus we may avoid all the heretofore frustrating factors of uninspired patron guidance of our work such as the patron's supine concessions to the nonsynergetical thinking, and therefore ignorantly conditioned reflexes, of the least well advised of the potential mass customers.

Typical of the subsidiary problems within the whole human survival problem, whose ramifications now go beyond the prerogatives of planners and must be solved, is the problem of pollution in general–pollution not only of our air and water but also of the information stored in our brains. We will soon have to rename our planet "Poluto." In respect to our planet's life sustaining atmosphere we find that, yes, we do have technically feasible ways of precipitating the fumes, and after this we say,

"But it costs too much." There are also ways of desalinating sea water, and we say, "But it costs too much." This too narrow treatment of the problem never faces the inexorably-evolving and solution-insistent problem of what it will cost when we don't have the air and water with which to survive. It takes months to starve to death, weeks to thirst to death, but only minutes to suffocate. We cannot survive without water for the length of time it takes to produce and install desalinization equipment adequate to supply, for instance, all of New York City. A sustained, and often threatened, water shortage in New York City could mean death for millions of humans. Each time the threat passes the old statement "it costs too much" again blocks realization of the desalinization capability.

Anybody who has been in Washington (and approximately everyone else everywhere today) is familiar with governmental budgeting and with the modes of developing public recognition of problems and of bringing about official determination to do something about solutions. In the end, the problems are rarely solved, not because we don't know how but because it is discovered either that it is said by those in authority that "it costs too much" or that when we identify the fundamental factors of the environmental problems—and laws are enacted to cope incisively with those factors—that there are no funds presently known to be available with which to implement the law. There comes a money bill a year later for implementation and with it the political criteria of assessing wealth by which the previous year's bill would now seemingly "cost too much." So compromises follow compromises. Frequently, nothing but political promises or underbudgeted solutions result. The original legislation partially stills the demands. The pressures on the politicians are lowered, and the lack of implementation is expeditiously shrugged off because of seemingly more pressing, seemingly higher priority, new demands for the money. The most pressing of those demands is for war, for which the politicians suddenly accredit weaponry acquisitions and military tasks costing many times their previously asserted concepts of what we can afford.

Thus under lethal emergencies vast new magnitudes of wealth come mysteriously into effective operation. We don't seem to be able to afford to do peacefully the logical things we say we ought to be doing to forestall warring—by producing enough to satisfy all the world needs. Under pressure we always find that we can afford to wage the wars brought about by the vital struggle of "have-nots" to share or take over the bounty of the "haves." Simply because it had seemed, theretofore, to cost too much to provide vital support of those "have-nots." The "haves" are thus forced in self-defense suddenly to articulate and realize productive wealth capabilities worth many times the amounts of monetary units they had known themselves to possess and, far more importantly, many times what it would have cost to give adequate economic support to the particular "have-nots" involved in the warring and, in fact, to all the world's "have-nots."

The adequately macro-comprehensive and micro-incisive solutions to any and all vital problems never cost too much. The production of heretofore nonexistent production tools and industrial networks of harnessed energy to do more work does not cost anything but human time which is refunded in time gained minutes after the inanimate machinery goes to work. Nothing is spent. Potential wealth has become real wealth. As it is clichéd "in the end" problem solutions always cost the least if paid for adequately at outset of the vital problem's recognition. Being vital, the problems are evolutionary, inexorable, and ultimately unavoidable by humanity. The constantly put-off or undermet costs and society's official bumbling of them clearly prove that man does not know at present what wealth is nor how much of whatever it may be is progressively available to him.

We have now flushed out a major variable in our general systems problem of man aboard Earth. The question "What is wealth?" commands our prime consideration.

The *Wall Street Journal* reported the September–October 1967 deliberations of the International Monetary Fund held at Rio De Janeiro, Brazil. Many years and millions of dollars were spent maneuvering for and assembling this monetary convention, and the net result was the weak opinion that it would soon be time to consider doing something about money. The convention felt our international balance of payments and its gold demand system to be inadequate. They decided that the old pirate's gold was still irreplaceable but that after a few years they might have to introduce some new "gimmick" to augment the gold as an international monetary base.

At present there is about seventy billion dollars of mined gold known to exist on board our Spaceship Earth. A little more than half of it–about forty billion–is classified as being "monetary"; that is, it exists in the forms of various national coinages or in the form of officially banked gold bullion bars. The remaining thirty billion is in private metallic hordes, jewelry, gold teeth, etc.

Since banks have no money of their own

and only our deposits on which they earn "interest," bank wealth or money consists only of accrued bank income. Income represents an average return of 5 per cent on capital invested. We may assume therefore from an estimate of the world's annual gross product that the capital assets, in the form of industrial production, on board our Spaceship Earth are at present worth in excess of a quadrillion dollars. The world's total of seventy billion dollars in gold represents only three one-thousandths of 1 per cent of the value of the world's organized industrial production resources. The gold supply is so negligible as to make it pure voodoo to attempt to valve the world's economic evolution traffic through the gold-sized needle's "eye."

Gold was used for trading by the Great Pirates in lieu of any good faith whatsoever—and in lieu of any mutual literacy, scientific knowledge, intelligence, or scientific and technical know-how on both sides of the trading. Gold trading assumed universal rascality to exist. Yet the realization of the planners' earnest conceptioning and feasible work on behalf of the ill-fated 60 per cent of humanity are entirely frustrated by this kind of nonsense.

We therefore proceed ever more earnestly with our general systems analysis of the problems of human survival, with the premise that at present neither the world's political officials nor its bankers know what wealth is. In organizing our thoughts to discover and clarify what wealth is we also will attempt to establish an effective means to develop immediate working procedures for solution of such big problems.

I have tried out the following intellectual filtering procedure with both multi-thousand general public audiences and with audiences of only a hundred or so advanced scholars and have never experienced disagreement with my progression of residual conclusions. I proceed as follows: I am going to make a series of analytical statements to you, and if anyone disagrees with me on any statement we will discard that statement. Only those of all my statements which remain 100 per cent unprotested will we rate as being acceptable to all of us.

First, I say, "No matter what you think wealth may be and no matter how much you have of it, you cannot alter one iota of yesterday." No protest? We've learned some lessons. We can say that wealth is irreversible in evolutionary processes. Is there anyone who disagrees with any of my statements thus far—about what wealth is or is not? Good—no disagreement—we will go on.

Now, I'm going to have a man in a shipwreck. He's rated as a very rich man, worth over a billion dollars by all of society's accredited conceptions of real wealth. He has taken with him on his voyage all his stocks and bonds, all his property deeds, all his check-books, and, to play it safe, has brought along a lot of diamonds and gold bullion. The ship burns and sinks, and there are no lifeboats, for they, too, have burned. If our billionaire holds on to his gold, he's going to sink a little faster than the others. So I would say he hadn't much left either of now or tomorrow in which to articulate his wealth, and since wealth cannot work backwardly his kind of wealth is vitally powerless. It is really a worthless pile of chips of an arbitrary game which we are playing and does not correspond to the accounting processes of our real universe's evolutionary transactions. Obviously the catastrophied bil-lionaire's kind of wealth has no control over either yesterday, now, or tomorrow. He can't extend his life with that kind of wealth unless he can persuade the one passenger who has a life-jacket to yield that only means of extending one life in exchange for one crazy moment's sense of possession of all the billionaire's sov-ereign-powers-backed legal tender, all of which the catastrophy-disillusioned and only moments earlier "powerfully rich" and now desperately helpless man would thankfully trade for the physical means of extending the years of his life; or of his wife.

It is also worth remembering that the validity of what our reputedly rich man in the shipwreck had in those real estate equities went back only to the validity "in the eyes of God" of the original muscle, cunning, and weapons-established–sovereign-claimed lands and their subsequent legal re-deedings as "legal" properties protected by the moral-or-no, weapons-enforced laws of the sovereign nations and their subsequent abstraction into limited-liability-corporation equities printed on paper stocks and bonds. The procedure we are pursuing is that of true democracy. Semi-democracy accepts the dictatorship of a majority in establishing its arbitrary, ergo, unnatural, laws. True democracy discovers by patient experiment and unanimous acknowledgement what the laws of nature or universe may be for the physical support and metaphysical satisfaction of the human intellect's function in universe.

I now go on to speculate that I think that what we all really mean by wealth is as follows: "Wealth is our organized capability to cope effectively with the environment in sustaining our healthy regeneration and decreasing both the physical and metaphysical restrictions of the forward days of our lives."

Is there any disagreement? Well, having first disposed of what wealth is not we now have produced a culled-out statement that roughly contains somewhere within it a precise definition of what wealth is. Now we can account *wealth* more precisely as *the number of forward days for a specific number of people we are physically prepared to sustain at a physically stated time and space liberating level of metabolic and metaphysical regeneration.*

We are getting sharper. Inasmuch as we now are learning more intimately about our Spaceship Earth and its radiation supply ship Sun on the one hand and on the other its Moon acting as the Earth's gravitationally pulsing "alternator" which together constitute the prime generator and regenerator of our life supporting system, I must observe also that we're not going to sustain life at all except by our successful impoundment of more of the Sun's radiant energy aboard our spaceship than we are losing from Earth in the energies of radiation or outwardly rocketed physical matter. We could burn up the Spaceship Earth itself to provide energy, but that would give us very little future. Our space vehicle is similar to a human child. It is an increasing aggregate of physical and metaphysical processes in contradistinction to a withering, decomposing corpse.

It is obvious that the real wealth of life aboard our planet is a forwardly-operative, metabolic, and intellectual regenerating system. Quite clearly we have vast amounts of income wealth as Sun radiation and Moon gravity to implement our forward success. Wherefore living only on our energy savings by burning up the fossil fuels which took billions of years to impound from the Sun or living on our capital by burning up our Earth's atoms is lethally ignorant and also utterly irresponsible to our coming generations and their forward days. Our children and their children are our future days. If we do not comprehend and realize our potential ability to support all life forever we are cosmicly bankrupt.

Having identified the ignorance of society regarding its wealth capabilities to be a major factor in the frustration of effective planning and having roughly identified the meaning of wealth to which everyone can realistically subscribe, and intending later to sharpen its identity, we will now tackle the next phase of humanity's total survival, prosperity, happiness, and regenerative inspiration with the problem-solving power of General Systems Theory as combined with both computer strategy—which is known as cybernetics—and with synergetics—the latter consisting of the solving of problems by starting with known behaviors of whole systems plus the known behaviors of some of the systems' parts, which advantageous information makes possible the discovery of other parts of the system and their respective behaviors, as for instance in geometry the known sum–180 degrees–of a triangle's angles, plus the known behavior of any two sides and their included angle and vice versa, enables the discovery and use of the precise values of the other three parts.

Synergetics discloses that wealth, which represents our ability to deal successfully with our forward energetic regeneration and provide increased degrees of freedom of initiation and noninterfering actions, breaks down cybernetically into two main parts: physical energy and metaphysical know-how. Physical energy in turn divides into two interchangeable phases: associative and disassociative–energy associative as matter and energy disassociative as radiation.

Stating first that physical universe is all energy and symbolizing energy by *E*, Einstein formulated his famous equation $E = M$ (matter's mass, explained in the terms of C^2–speed of an omni-directional [radiant] surface wave's expansion, unfettered, in a vacuum). Energy as matter and energy as radiation, as Einstein had generalized hypothetically, were explicitly proven by fission to be interchangeable covariants.

Physicists also have discovered experimentally that energy can neither be exhausted nor originated. Energy is finite and infinitely conserved. This experimentally proven realization of some prime facts of physical universe contradicts the thoughts of cosmologists, cosmogonists, and society's economics expressed before the speed of light was measured at the beginning of the twentieth century.

I came to Harvard University in the beginning of the century—just before World War I. At that time, it was as yet the consensus of scholarly thinking that because the universe itself was seemingly a system it, too, must be subject to entropy, by which every (local) system was found experimentally to be continually losing energy. Hence, the universe itself was thought to be losing energy. This indicated that the universe was "running down," at which time evolution would abandon its abnormal energetic behavior and all would return to Newton's norm of "at rest." This being so, it was also assumed that all those who expended energy were recklessly speeding the end. This was the basis of yesterday's conservatism. All who expended energy in bringing about further evolutionary changes were to be abhorred. They were to be known as reckless spenders.

All this was assumed to be true before experiments at the beginning of the twentieth century gave scientists knowledge of the speed of light and of radiation in general. Thus, we suddenly discovered it took eight minutes for the light to get to us from the sun, two and one-half years from the next nearest star beyond the sun, and many years for it to reach us from other stars. We learned only two-thirds of a century ago that many stars we considered as *instantly* there had burned out thousands of years ago. Universe is not simultaneous.

Then Einstein, Planck, and other leading scientists said, "We're going to have to reassess and redefine the physical universe." They defined the physical universe as "an aggregate of non-simultaneous and only partially overlapping transformation events." And they then said: "We must discover what it is we see when we observe new life forming. It could be that when energy disassociates here it always may be reassociating somewhere else." And that in all subsequent experimentation proved to be the case. The scientists found that the energy redistributions always added up to 100 per cent. The scientists then formulated a new description of the physical universe which they called the new *"law of conservation of energy,"* which said that "physical experiments disclose that energy can neither be created nor lost." Energy is not only conserved but it is also finite. It is a closed system. The universe is a mammoth perpetual motion process. We then see that the part of our wealth which is physical energy is conserved. It cannot be exhausted, cannot be spent, which means exhausted. We realize that the word "spending" is now scientifically meaningless and is therefore obsolete.

I referred earlier to man's discovery of the lever. Having used levers for millenniums, man thought of taking a series of bucket-ended levers and of inserting the nonbucket end of each lever perpendicularly into a shaft—arranged one after another like spokes around a wheel. He mounted that shaft in bearings, put it under a waterfall, and let gravity fill each bucket in turn and then pull each filled bucket toward the center of the planet Earth, thus progressively rotating all the levers and making the wheel and its shaft rotate with great power. Then man connected this revolving shaft by pulley belts to other pulleys on other shafts which drove machines to do many metabolically-regenerating tasks in a way unmatchable by man's muscle power alone. Man thus began for the first time to really employ his intellect in the most important way. He discovered how to use energy as matter in the form of levers, shafts, gear trains, and dams, and how to take advantage of and use the energy as Sun radiation which vaporized and elevated water as atmospheric cloud, allowing it then to be precipitated and pulled back toward the center of the spherical Earth from the spherical mantle of clouds in the form of water molecules collected in droplets. From this moment of comprehending energy circuits, and thenceforth, man's really important function in universe was his intellection, which taught him to intercept and redirect local energy patternings in universe and thus to reorganize and shunt those flow patterns so that they would impinge on levers to increase humanity's capabilities to do the manifold tasks leading directly and indirectly toward humanity's forward metabolic regeneration.

What we now have demonstrated metaphysically is that every time man makes a new experiment he always learns more. He cannot learn less. He may learn that what he thought was true was not true. By the elimination of a false premise, his basic capital wealth which is his given lifetime is disembarrassed of further preoccupation with considerations of how to employ a worthless time-consuming hypothesis. Freeing his time for its more effective exploratory investment is to give man increased wealth.

We find experimentally, regarding the metaphysical phenomenon, intellect, which we call know-how, that every time we employ and test our intellectual know-how by experimental rearrangement of physical energy interactions (either associated as mass or disassociated as radiation, free energy) we always learn more. The know-how can only increase. Very interesting. Now we have carefully examined and experimented with the two basic constituents of wealth–the physical and the metaphysical.

Sumtotally, we find that the physical constituent of wealth–energy–cannot decrease and that the metaphysical constituent–know-how–can only increase. This is to say that everytime we use our wealth it increases. This is to say that, countering entropy, wealth can only increase. Whereas entropy is increasing disorder evoked by dispersion of energy, wealth locally is increased order–that is to say, the increasingly orderly concentration of physical power in our ever-expanding locally explored and comprehended universe by the metaphysical capability of man, as informed by repeated experiences from which he happens in an unscheduled manner to progressively distill the ever-increasing inventory of omni-interrelated and omni-interaccommodative generalized principles found to be operative in all the special-case experiences. Irreversible wealth is the so far attained effective magnitude of our physically organized ordering of the use of those generalized principles.

Wealth is anti-entropy at a most exquisite

degree of concentration. The difference between mind and brain is that brain deals only with memorized, subjective, special-case experiences and objective experiments, while mind extracts and employs the generalized principles and integrates and interrelates their effective employment. Brain deals exclusively with the physical, and mind exclusively with the metaphysical. Wealth is the product of the progressive mastery of matter by mind, and is specifically accountable in forward man-days of established metabolic regeneration advantages spelt out in hours of life for specific numbers of individuals released from formerly prescribed entropy-preoccupying tasks for their respectively individual yet inherently co-operative elective investment in further anti-entropic effectiveness.

Because our wealth is continually multiplying in vast degree unbeknownst and unacknowledged formally by human society, our economic accounting systems are unrealistically identifying wealth only as matter and are entering know-how on the books only as salary liabilities; therefore, all that we are discovering mutually here regarding the true nature of wealth comes as a complete surprise to world society—to both communism and to capitalism alike. Both social co-operation and individual enterprise interact to produce increasing wealth, all unrecognized by ignorantly assumed lethally competitive systems. All our formal accounting is anti-synergetic, depreciative, and entropic mortgagization, meaning death by inversally compounding interest. Wealth as anti-entropy developes compound interest through synergy, which growth is as yet entirely unaccounted anywhere around Earth in any of its political economic systems. We give an intrinsic value to the material. To this we add the costs of manufacturing which include energy, labor, overhead, and profit. We then start depreciating this figure assuming swift obsolescence of the value of the product. With the exception of small royalties, which are usually avoided, no value is given for the inventiveness or for the synergistic value given by one product to another by virtue of their complementarity as team components whose teamwork produces results of enormous advantage, as for instance do the invention of alloyed drill bits of oil rigs bring petroleum from nonuse to use.

As a consequence of true wealth's unaccounted, inexorably synergistic multiplication of ever-increasing numbers of humanity's ever-increasing forward days, in this century alone we have gone from less than 1 per cent of humanity being able to survive in any important kind of health and comfort to 44 per cent of humanity surviving at a standard of living unexperienced or undreamed of before. This utterly unpredicted synergistic success occurred within only two-thirds of a century despite continually decreasing metallic resources per each world person. It happened without being consciously and specifically attempted by any government or business. It also happened only as a consequence of man's inadvertently becoming equipped synergistically to do progressively more with less.

As we have learned, synergy is the only word in our language which identifies the meaning for which it stands. Since the word is unknown to the average public, as I have already pointed out, it is not at all surprising that synergy has not been included in the economic accounting of our wealth transactions or in assessing our common wealth capabilities. The synergetic aspect of industry's doing ever more work with ever less investment of time and energy per each unit of performance of each and every function of the weapons carriers of the sea, air, and outer space has never been formally accounted as a capital gain of land-situated society. The synergistic effectiveness of a world-around integrated industrial process is inherently vastly greater than the confined synergistic effect of sovereignly operating separate systems. Ergo, only complete world desovereignization can permit the realization of an all humanity high standard support. But the scientific facts are that simple tools that make complex tools are synergetically augmented by progressively more effective and previously unpredicted chemical elements alloying. The whole history of world industrialization demonstrates constantly surprising new capabilities resulting from various synergetic interactions amongst both the family of ninety-two regenerative and the trans-uranium members of the uniquely-behaving chemical elements family.

Complex environmental evolution is synergetically produced by the biologicals and their tools as well as by the great inanimate physiological complex of events such as earthquakes and storms which have constant challenging effect upon the individual biological inventiveness wherein both the challenges and the cause are regenerative. Our common wealth is also multiplied in further degree by experimentally-derived information which is both multiplying and integrating the wealth advantage at an exponential rate. The synergetic effect upon the rate of growth of our incipient world common wealth augmentation has been entirely overlooked throughout all the accounting systems of all the ideologically-divergent political systems. Our wealth is inherently common wealth and our common wealth can only increase, and it is increasing at a constantly self-accelerating synergetic rate.

However, we inadvertently dip into our real, unaccountedly fabulous wealth in a very meager way only when our political leaders become scared enough by the challenges of an impressively threatening enemy. Then only do socialism and capitalism alike find that they have to afford whatever they need. The only limitation in the realization of further wealth is that production engineers must be able to envision and reduce to design and practice the production-multiplying steps to be taken, which progressive envisioning depends both on the individual and on the experimentally proven, but as yet untapped, state of the pertinent metaphysical arts, as well as upon the range of resources strategically available at the time and in particular upon the inventory of as yet unemployed but relevant inventions.

In respect to physical resources, until recently man had assumed that he could produce his buildings, machinery, and other products only out of the known materials. From time to time in the past, scientists discovered new alloys which changed the production engineering prospects. But now in the aerospace technology man has developed his metaphysical capabilities to so advanced a degree that he is evolving utterly unique materials "on order." Those new materials satisfy the prespecified physical behavior characteristics which transcend those of any substance previously known to exist anywhere in the universe. Thus, the re-entry nosecones of the man-launched and rocketed satellites were developed. Synergy is of the essence. Only under the stresses of total social emergencies as thus far demonstrated by man do the effectively adequate alternative technical strategies synergetically emerge. Here we witness mind over matter and humanity's escape from the limitations of his exclusive identity only with some sovereignized circumscribed geographical locality.

7 *integral functions*

THE first census of population in the United States was taken in 1790. In 1810 the United States Treasury conducted the first economic census of the young democracy. There were at that time one million families in this country. There were also one million human slaves. This did not mean that each family had a human slave; far from it. The slaves were owned by relatively few.

The Treasury adjudged the monetary value of the average American homestead, lands, buildings, furnishings, and tools to be worth sumtotally $350 per family. The Treasury appraised the average worth of each slave as $400. It was estimated that the wilderness hinterlands of America were worth $1,500 per family. The foregoing assets plus the canals and toll roads brought the equity of each family to a total of $3,000. This made the national wealth of the United States, as recognized by man, worth three billion dollars.

Let us assume that, practicing supreme wisdom, the united American citizens of 1810 had convened their most reliably esteemed and farsighted leaders and had asked them to undertake a 150-year, grand economic and technical plan for most effectively and swiftly developing America's and the world's life-support system—to be fully realized by 1960. At that time, it must be remembered, the telegraph had not been invented. There were no electromagnetics or mass-produced steel. Railroads were as yet undreamed of, let alone wireless, X-ray, electric light, power by wire, and electric motors. There was no conception of the periodic table of the atoms or of the existence of an electron. Had any of our forefathers committed our wealth of 1810 toward bouncing radar impulses off the Moon he would have been placed in a lunatic asylum.

Under those 1810 circumstances of an assumed capital wealth of the united American states, both public and private, amounting to only three billion dollars, it is preposterous to think of humanity's most brilliant and powerful leaders electing to invest their "all" of three billion dollars in a "thousand times more expensive" ten-trillion-dollar adventure such, however, as has since transpired, but only under the war-enforced threat of disintegration of the meager rights won thus far by common man from history-long tyrannical powers of a techno-illiterate and often cruel few.

In 1810 it was also unthinkable by even the most brilliant leaders of humanity that 160 years hence, in 1970, the gross national product of the United States would reach one trillion dollars per year. (This is to be compared with the meager forty billion of the world's total monetary gold supply.) Assuming a 10 per cent rate of earnings, this 1970 trillion-dollar product would mean that a capital base of ten trillion dollars was operative within the United States alone where the 1810 national leaders had accredited only three billion dollars of national assets. The wisest humans recognized in 1810 only one three-hundredth of 1 per cent of the immediately thereafter "proven value" of the United States' share of the world's wealth-generating potentials. Of course, those wisest men of the times would have seen little they could afford to do.

Our most reliable, visionary, and well-informed great-grandfathers of 1810 could not have foreseen that in the meager century and one-half of all the billionsfold greater reaches of known universal time that human life-span would be trebled, that the yearly real income of the individual would be tenfolded, that the majority of diseases would be banished, and human freedom of realized travel onehundred-folded; that humans would be able to whisper effortlessly in one another's ear from anywhere around the world apart and at a speed of seven hundred million miles an hour, their audibility clearly reaching to the planet Venus; and that human vision around Earth's spherical deck would be increased to see local pebbles and grains of sand on the moon.

Now in 1969, 99.9 per cent of the accelerating accelerations of the physical environment changes effecting all humanity's evolution are transpiring in the realms of the electromagnetic spectrum realities which are undetectable directly by the human senses. Because they are gestating invisibly it is approximately impossible for world society to comprehend that the changes in the next thirty-five years—ushering in the twenty-first century—will be far greater than in our just completed century and one-half since the first United States economic census. We are engulfed in an invisible tidal wave which, as it draws away, will leave humanity, if it survives, cast up upon an island of universal success uncomprehending how it has all happened.

But we can scientifically assume that by the twenty-first century either humanity will not be living aboard Spaceship Earth or, if approximately our present numbers as yet remain aboard, that humanity then will have recognized and organized itself to realize effectively the fact that humanity can afford to do anything it needs and wishes to do and that it cannot afford anything else. As a consequence Earth-planet-based humanity will be physically and economically successful and individually free in the most important sense. While all enjoy total Earth no human will be interfering with the other, and none will be profiting at the expense of the other. Humans will be free in the sense that 99.9 per cent of their waking hours will be freely investable at their own discretion. They will be free in the sense that they will not struggle for survival on a "you" or "me" basis, and will therefore be able to trust one another and be free to co-operate in spontaneous and logical ways.

It is also probable that during that one-third of a century of the curtain raising of the twenty-first century that the number of boo-boo's, biased blunders, short-sighted misjudgments, opinionated self-deceits of humanity will total, at minimum, six hundred trillion errors. Clearly, man will have backed into his future while evolution, operating as inexorably as fertilized ovaries gestate in the womb, will have brought about his success in ways as synergetically unforeseeable to us today as were the ten-trillion-dollar developments of the last 150 years unforeseen by our wisest great-grandfathers of 1810.

All of this does not add up to say that man is stupidly ignorant and does not deserve to prosper. It adds up to the realization that in the design of universal evolution man was given an enormous safety factor as an economic cushion, within which to learn by trial and error to dare to use his most sensitively intuited intellectual conceptioning and greatest vision in joining forces with all of humanity to advance into the future in full accreditation of the individual human intellect's most powerfully loving conceptions of the potential functioning of man in universe. All the foregoing is to say also that the opinions of any negatively conditioned reflexes regarding what I am saying and am about to say are unrealistically inconsequential.

I have so far introduced to you a whole new synergetic assessment of wealth and have asked that you indicate your disagreement if you detected fallacies in the progressively-stated concepts of our common wealth. Thus we have discovered together that we are unanimous in saying that we can afford to do anything we need or wish to do.

It is utterly clear to me that the highest priority need of world society at the present moment is a realistic economic accounting system which will rectify, for instance, such nonsense as the fact that a top toolmaker in India, the highest paid of all craftsmen, gets only as much per month for his work in India as he could earn per day for the same work if he were employed in Detroit, Michigan. How can India develop a favorable trade balance under those circumstances? If it can't have a workable, let alone favorable balance, how can these half-billion people participate in world intercourse? Millions of Hindus have never heard of America, let alone the international monetary system. Said Kipling "East is east and west is west and never the twain shall meet."

As a consequence of the Great Pirates' robbing Indo-China for centuries and cashing in their booty in Europe, so abysmally impoverished, underfed and physically afflicted have India's and Ceylon's billions of humans been throughout so many centuries that it is their religious belief that life on Earth is meant to be exclusively a hellish trial and that the worse the conditions encountered by the individual the quicker his entry into heaven. For this reason attempts to help India in any realistic way are looked upon by a vast number of India's population as an attempt to prevent their entry into heaven. All this because they have had no other way to explain life's hopelessness. On the other hand, they are extremely capable thinkers, and free intercourse with the world could change their views and fate. It is paradoxical that India's population should starve as one beef cattle for every three people wander through India's streets, blocking traffic as sacred symbols of nonsense. Probably some earlier conquerors intent to reserve the animals for their exclusive consumption as did later the kings of European nations decreed that God had informed the king that he alone was to eat animal meat and therefore God forbade the common people under penalty of death from killing a beef cattle for their own consumption.

One of the myths of the moment suggest that wealth comes from individual bankers and capitalists. This concept is manifest in the myriad of charities that have to beg for alms for the poor, disabled, and helpless young and old in general. These charities are a hold-over from the old pirate days, when it was thought that there would never be enough to go around. They also are necessitated by our working assumption that we cannot afford to take care of all the helpless ones. Counselled by our bankers, our politicians say we can't afford the warring and the great society, too. And because of the mythical concept that the wealth which is disbursed is coming from some magically-secret private source, no free and healthy individual wants that "hand out" from the other man, whoever he may be. Nor does the individual wish to be on the publicly degrading "dole" line.

After World War II several million of our well-trained, healthiest young people came suddenly out of the military service. Because we had automated during the war to a very considerable degree to meet the "war challenges" there were but few jobs to offer them. Our society could not say realistically that the millions of their healthiest, best informed young were unfit because they couldn't get a job, which had until that historical moment been the criteria of demonstrated fitness in Darwin's "survival only of the fittest" struggle. In that emergency we legislated the GI Bill and sent them all to schools, colleges, and universities. This act was politically rationalized as a humanly dignified fellowship reward of their war service and not as a "hand out." It produced billions of dollars of new wealth through the increased know-how and intelligence thus released, which synergetically augmented the spontaneous initiative of that younger generation. In legislating this "reckless spending" of wealth we didn't know that we had produced a synergetic condition that would and did open the greatest prosperity humanity has ever known.

Through all pre-twentieth-century history wars were devastating to both winners and losers. The pre-industrial wars took the men from the fields, and the fields where the exclusively agricultural-wealth germinated, were devastated. It came as a complete surprise, therefore, that the first World War, which was the first full-fledged industrial-era war, ended with the United States in particular but Germany, England, France, Belgium, Italy, Japan, and Russia in lesser degree all coming out of the war with much greater industrial production capabilities than those with which they had entered. That wealth was soon misguidedly invested in the second World War, from which all the industrial countries emerged with even greater wealth producing capabilities, despite the superficial knockdown of the already obsolete buildings. It was irrefutably proven that the destruction of the buildings by bombing, shell fire, and flames left the machinery almost unharmed. The productive tooling capabilities multiplied unchecked, as did their value.

This unexpected increase in wealth by industrial world wars was caused by several facts, but most prominently by the fact that in the progressive acquisition of instruments and tools which produce the even more effective complex of industrial tools, the number of special purpose tools that made the end-product armaments and ammunition was negligible in comparison with the redirectable productivity of the majority of the general-purpose tools that constituted the synergistic tool complex. Second, the wars destroyed the obsolete tool-enclosing brick-and-wood structures whose factual availability, despite their obsolescence, had persuaded their owners to over extend the structures' usefulness and exploitability. This drive to keep milking the old proven cow not risking the production of new cows had blocked the acquisition of up-to-date tools. Third, there was the synergetic surprise of alternative or "substitute" technologies which were developed to bypass destroyed facilities. The latter often proved to be more efficient than the tools that were destroyed. Fourth, the metals themselves not only were not destroyed but were acceleratingly reinvested in new, vastly higher-performance per pound tools. It was thus that the world war losers such as Germany and Japan became overnight the postwar industrial winners. Their success documented the fallacy of the whole economic evaluation system now extant.

Thus again we see that, through gradually increasing use of his intuition and intellect, man has discovered many of the generalized principles that are operative in the universe and has employed them objectively but separately in extending his internal metabolic regeneration by his invented and detached tool extensions and their remote operation affected by harnessing inanimate energy. Instead of trying to survive only with his integral set of tool capabilities—his hands—to pour water into his mouth, he invents a more effective wooden, stone, or ceramic vessel so that he not only can drink from it but carry water with him and extend his hunting and berry picking. All tools are externalizations of originally integral functions. But in developing each tool man also extends the limits of its usefulness, since he can make bigger cups hold liquids too hot or chemically destructive for his hands. Tools do not introduce new principles but they greatly extend the range of conditions under which the discovered control principle may be effectively employed by man. There is nothing new in world technology's growth. It is only the vast increase of its effective ranges that are startling man. The computer is an imitation human brain. There is nothing new about it, but its capacity, speed of operation, and tirelessness, as well as its ability to operate under environmental conditions intolerable to the human anatomy, make it far more effective in performing special tasks than is the skull and tissue encased human brain, minus the computer.

What is really unique about man is the magnitude to which he has detached, deployed, amplified, and made more incisive all of his many organic functionings. Man is unique among all the living phenomena as the most adaptable omni-environment penetrating, exploring, and operating organism being initially equipped to invent intellectually and self-disciplined, dexterously, to make the tools with which thus to extend himself. The bird, the fish, the tree are all specialized, and their special capability-functioning tools are attached integrally with their bodies, making them incapable of penetrating hostile environments. Man externalizes, separates out, and increases each of his specialized function capabilities by inventing tools as soon as he discovers the need through oft-repeated experiences with unfriendly environmental challenges. Thus, man only temporarily employs his integral equipment as a specialist, and soon shifts that function to detached tools. Man cannot compete physically as a muscle and brained automaton

—as a machine—against the automated power tools which he can invent while metaphysically mastering the energy income from universe with which evermore powerfully to actuate these evermore precise mass-production tools. What man has done is to decentralize his functions into a world-around-energy-networked complex of tools which altogether constitute what we refer to as world industrialization.

8 *the regenerative landscape*

THUS man has developed an externalized metabolic regeneration organism involving the whole of Spaceship Earth and all its resources. Any human being can physically employ that organism, whereas only one human can employ the organically integral craft tool. All 91 of the 92 chemical elements thus far found aboard our spaceship are completely involved in the world-around industrial network. The full family of chemical elements is unevenly distributed, and therefore our total planet is at all times involved in the industrial integration of the unique physical behaviors of each of all the elements. Paradoxically, at the present moment our Spaceship Earth is in the perilous condition of having the Russians sitting at one set of the co-pilot's flying controls while the Americans sit at the other. France controls the starboard engines, and the Chinese control the port engines, while the United Nations controls the passenger operation. The result is an increasing number of U. F. O. hallucinations of sovereign states darting backwards and forwards and around in circles, getting nowhere, at an incredibly accelerating rate of speed.

All of humanity's tool extensions are divisible into two main groups: the craft and the industrial tools. I define the craft tools as all those tools which could be invented by one man starting all alone, naked in the wilderness, using only his own experience and his own integral facilities. Under these isolated conditions he could and did invent spears, slings, bows, and arrows, etc. By industrial tools I mean all the tools that cannot be produced by one man, as for instance the S.S. *Queen Mary*. With this definition, we find that the spoken word, which took a minimum of two humans to develop, was the first industrial tool. It brought about the progressive integration of all individual generation-to-generation experiences and thoughts of all humanity everywhere and everywhen. The Bible says, "In the beginning was the word"; I say to you, "In the beginning of industrialization was the spoken word." With the graphic writing of the words and ideas we have the beginning of the computer, for the computer *stores* and retrieves information. The written word, dictionary and the book were the first information storing and retrieving systems.

The craft tools are used initially by man to make the first industrial tools. Man is using his hands today most informatively and expertly only to press the buttons that set in action the further action of the tools which reproduce other tools which may be used informatively to make other tools. In the craft economies craftsman artists make only end- or consumer-products. In the industrial economy the craftsman artists make the tools and the tools make the end- or consumer-products. In this industrial development the mechanical advantages of men are pyramided rapidly and synergetically into invisible magnitudes of ever more incisive and inclusive tooling which produces ever more with ever less resource investment per each unit of end-product, or service, performance.

As we study industrialization, we see that we cannot have mass production unless we have mass consumption. This was effected evolutionarily by the great social struggles of labor to increase wages and spread the benefits and prevent the reduction of the numbers of workers employed. The labor movement made possible mass purchasing; ergo, mass production; ergo, low prices on vastly improved products and services, which have altogether established entirely new and higher standards of humanity's living.

Our labor world and all salaried workers, including school teachers and college professors, are now, at least subconsciously if not consciously, afraid that automation will take away their jobs. They are afraid they won't be able to do what is called "earning a living," which is short for earning the right to live. This term implies that normally we are supposed to die prematurely and that it is abnormal to be able to earn a living. It is paradoxical that only the abnormal or exceptional are entitled to prosper. Yesterday the term even inferred that success was so very abnormal that only divinely ordained kings and nobles were entitled to eat fairly regularly.

It is easy to demonstrate to those who will take the time and the trouble to unbias their thoughts that automation swiftly can multiply the physical energy part of wealth much more rapidly and profusely than can man's muscle and brain-reflexed–manually-controlled production. On the other hand humans alone can foresee, integrate, and anticipate the new tasks to be done by the progressively automated wealth-producing machinery. To take advantage of the fabulous magnitudes of real wealth waiting to be employed intelligently by humans and unblock automation's postponement by organized labor we must give each human who is or becomes unemployed a life fellowship in research and development or in just simple thinking. Man must be able to dare to think truthfully and to act accordingly without fear of losing his franchise to live. The use of mind fellowships will permit humans comprehensively to expand and accelerate scientific exploration and experimental prototype development. For every 100,000 employed in research and development, or just plain thinking, one probably will make a breakthrough that will more than pay for the other 99,999 fellowships. Thus, production will no longer be impeded by humans trying to do what machines can do better. Contrariwise, omni-automated and inanimately powered production will unleash humanity's unique capability—its metaphysical capability. Historically speaking, these steps will be taken within the next decade. There is no doubt about it. But not without much social crisis and consequent educational experience and discovery concerning the nature of our unlimited wealth.

Through the universal research and development fellowships, we're going to start emancipating humanity from being muscle and reflex machines. We're going to give everybody a chance to develop their most powerful mental and intuitive faculties. Given their research and development fellowship, many who have been frustrated during their younger years may feel like going fishing. Fishing provides an excellent opportunity to think clearly; to review one's life; to recall one's earlier frustrated and abandoned longings and curiosities. What we want everybody to do is to *think* clearly.

We soon will begin to generate wealth so rapidly that we can do very great things. I would like you to think what this may do realistically for living without spoiling the landscape, or the antiquities or the trails of humanity throughout the ages, or despoiling the integrity of romance, vision, and harmonic creativity. All the great office buildings will be emptied of earned living workers, and the automated office-processing of information will be centralized in the basements of a few buildings. This will permit all the modernly mechanized office buildings to be used as dwelling facilities.

When we approach our problems on a universal, general systems basis and progressively eliminate the irrelevancies, somewhat as we peel petals from an artichoke, at each move we leave in full visibility the next most important layer of factors with which we must deal. We gradually uncover *you* and *me* in the heart of now. But evolution requires that we comprehend each layer in order to unpeel it. We have now updated our definitions of universe by conforming them with the most recent and erudite scientific findings such as those of Einstein and Planck. Earlier in our thinking we discovered man's function in universe to be that of the most effective metaphysical capability experimentally evidenced thus far within our locally observable phases and time zones of universe. We have also discovered that it is humanity's task to comprehend and set in order the special case facts of human experience and to win therefrom knowledge of the a priori existence of a complex of generalized, abstract principles which apparently altogether govern all physically evolving phenomena of universe.

We have learned that only and exclusively through use of his mind can man inventively employ the generalized principles further to conserve the locally available physical energy of the only universally unlimited supply. Only thus can man put to orderly advantage the various, local, and otherwise disorderly behaviors of the entropic, physical universe. Man can and may metaphysically comprehend, anticipate, shunt, and meteringly introduce the evolutionarily organized environment events in the magnitudes and frequencies that best synchronize with the patterns of his successful and metaphysical metabolic regeneration while ever increasing the degrees of humanity's space and time freedoms from yesterday's ignorance sustaining survival procedure chores and their personal time capital wasting.

Now we have comprehended and peeled off the layers of petals which disclosed not only that physical energy is conserved but also that it is ever increasingly deposited as a fossil-fuel savings account aboard our Spaceship Earth through photosynthesis and progressive, complex, topsoil fossilization buried ever deeper within Earth's crust by frost, wind, flood, volcanoes, and earthquake upheavals. We have thus discovered also that we can make all of humanity successful through science's world-engulfing industrial evolution provided that we are not so foolish as to continue to exhaust in a split second of astronomical history the orderly energy savings of billions of years' energy conservation aboard our Spaceship Earth. These energy savings have been put into our Spaceship's life–regeneration-guaranteeing bank account for use only in self-starter functions.

The fossil fuel deposits of our Spaceship Earth correspond to our automobile's storage battery which must be conserved to turn over our main engine's self-starter. Thereafter, our "main engine," the life regenerating processes, must operate exclusively on our vast daily energy income from the powers of wind, tide, water, and the direct Sun radiation energy. The fossil-fuel savings account has been put aboard Spaceship Earth for the exclusive function of getting the new machinery built with which to support life and humanity at ever more effective standards of vital physical energy and reinspiring metaphysical sustenance to be sustained exclusively on our Sun radiation's and Moon pull gravity's tidal, wind, and rainfall generated pulsating and therefore harnessable energies. The daily income energies are excessively adequate for the operation of our main industrial engines and their automated productions. The energy expended in one minute of a tropical hurricane equals the combined energy of all the U.S.A. and U.S.S.R. nuclear weapons. Only by understanding this scheme may we continue for all time ahead to enjoy and explore universe as we progressively harness evermore of the celestially generated tidal and storm generated wind, water, and electrical power concentrations. We cannot afford to expend our fossil fuels faster than we are "recharging our battery," which means precisely the rate at which the fossil fuels are being continually deposited within Earth's spherical crust.

We have discovered that it is highly feasible for all the human passengers aboard Spaceship Earth to enjoy the whole ship without any individual interfering with another and without any individual being advantaged at the expense of another, provided that we are not so foolish as to burn up our ship and its operating equipment by powering our prime operations exclusively on atomic reactor generated energy. The too-shortsighted and debilitating exploitation of fossil fuels and atomic energy are similar to running our automobiles only on the self-starters and batteries and as the latter become exhausted replenishing the batteries only by starting the chain reaction consumption of the atoms with which the automobiles are constituted.

We have discovered also why we were given our intellectual faculties and physical extension facilities. We have discovered that we have the inherent capability and inferentially the responsibility of making humanity comprehensively and sustainably successful. We have learned the difference between brain and mind capabilities. We have learned of the superstitions and inferiority complexes built into all humanity through all of history's yesterdays of slavish survival under conditions of abysmal illiteracy and ignorance wherein only the most ruthless, shrewd, and eventually brutish could sustain existence, and then for no more than a third of its known potential life span.

This all brings us to a realization of the enormous educational task which must be successfully accomplished right now in a hurry in order to convert man's spin-dive toward oblivion into an intellectually mastered power pull-out into safe and level flight of physical and metaphysical success, whereafter he may turn his Spaceship Earth's occupancy into a universe exploring advantage. If it comprehends and reacts effectively, humanity will open an entirely new chapter of the experiences and the thoughts and drives thereby stimulated.

Most importantly we have learned that from here on it is success for all or for none, for it is experimentally proven by physics that "unity is plural and at minimum two"—the complementary but not mirror-imaged proton and neutron. You and I are inherently different and complementary. Together we average as zero--that is, as eternity.

Now having attained that cosmic degree of orbital conceptioning we will use our retro-rocket controls to negotiate our reentry of our

Spaceship Earth's atmosphere and return to our omni-befuddled present. Here we find ourselves maintaining the fiction that our crossbreeding World Man consists fundamentally of innately different nations and races which are the antithesis of that crossbreeding. Nations are products of many generations of local inbreeding in a myriad of remote human enclaves. With grandfather chiefs often marrying incestuously the gene concentrations brought about hybrid nationally-unique physiological characteristics which in the extreme northern hibernations bleached out the human skin and in the equatorial casting off of all clothing inbred darkly tanned pigmentation. All are the consequence only of unique local environment conditions and super inbreeding.

The crossbreeding world people on the North American continent consist of two separate *input* sets. The first era input set consists of those who came with the prevailing winds and ocean currents eastward to the North, South, and Central Americas by raft and by boat from across the Pacific, primarily during an age which started at least thirty thousand years ago, possibly millions of years ago, and terminated three hundred years ago. The eastbound trans-Pacific migration peopled the west coasts of both South and North America and migrated inland towards the two continents' middle ground in Central America and Mexico. In Mexico today will be found every type of human characteristic and every known physiognomy, each of which occur in such a variety of skin shades from black to white that they do not permit the ignorance-invented "race" distinctions predicated only superficially on extreme limits of skin color. The second or westbound input era set of crossbreeding world man now peopling the Americas consists of the gradual and slower migration around the world from the Pacific Ocean westward into the wind, "following the sun," and travelling both by sea through Malaysia, across the Indian Ocean up the Persian Gulf into Mesopotamia and overland into the Mediterranean, up the Nile from East Africa into the South and North Atlantic to America—or over the Chinese, Mongolian, Siberian, and European hinterlands to the Atlantic and to the Americas.

Now both east and westbound era sets are crossbreeding with one another in ever-accelerating degree on America's continental middleground. This omni reintergration of world man from all the diverse hybrids is producing a crossbred people on the Pacific Coast of North America. Here with its aerospace and oceans penetrating capabilities, a world type of humanity is taking the springboard into all of the hitherto hostile environments of universe into the ocean depths and into the sky and all around the Earth.

Returning you again to our omni-befuddled present, we realize that reorganization of humanity's economic accounting system and its implementation of the total commonwealth capability by total world society, aided by the computer's vast memory and high speed recall comes first of all of the first-things-first that we must attend to to make our space vehicle Earth a successful man operation. We may now raise our sights, in fact must raise our sights, to take the initiative in planning the world-around industrial retooling revolution. We must undertake to increase the performance per pound of the world's resources until they provide all of humanity a high standard of living. We can no longer wait to see whose biased political system should prevail over the world.

You may not feel very confident about how you are going to earn your right to live under such world-around patronless conditions. But I say to you the sooner you do the better chance we have of pulling out of humanity's otherwise fatal nose dive into oblivion. As the world political economic emergencies increase, remember that we have discovered a way to make the total world work. It must be initiated and in strong momentum before we pass the point of no return. You may gain great confidence from the fact that your fellow men, some of them your great labor leaders, are already aware and eager to educate their own rank and file on the fallacy of opposition to automation.

I have visited more than three hundred universities and colleges around the world as an invited and appointed professor and have found an increasing number of students who understand all that we have been reviewing. They are comprehending increasingly that elimination of war can only be realized through a design and invention revolution. When it is realized by society that wealth is as much everybody's as is the air and sunlight, it no longer will be rated as a personal handout for anyone to accept a high standard of living in the form of an annual research and development fellowship.

I have owned successively, since boyhood, fifty-four automobiles. I will never own another. I have not given up driving. I began to leave my cars at airports—never or only infrequently getting back to them. My new pattern requires renting new cars at the airports as needed. I am progressively ceasing to own things, not on a political-schism basis, as for instance Henry George's ideology, but simply on a practical basis. Possession is becoming progressively burdensome and wasteful and therefore obsolete.

Why accumulate mementos of far away places when you are much more frequently in those places than at your yesterday's home, nation, state, city, and street identified residences, as required for passport, taxing, and voting functions? Why not completely restore the great cities and buildings of antiquity and send back to them all their fragmented treasures now deployed in the world's museums? Thus, may whole eras be reinhabited and experienced by an ever increasingly interested, well-informed, and inspired humanity. Thus, may all the world regain or retain its regenerative metaphysical mysteries.

I travel between Southern and Northern hemispheres and around the world so frequently that I no longer have any so-called normal winter and summer, nor normal night and day, for I fly in and out of the shaded or sun-flooded areas of the spinning, orbiting Earth with ever-increased frequency. I wear three watches to tell me what time it is at my "home" office, so that I can call them by long distance telephone. One is set for the time of day in the place to which I am next going, and one is set temporarily for the locality in which I happen to be. I now see the Earth realistically as a sphere and think of it as a spaceship. It is big, but it is comprehensible. I no longer think in terms of "weeks" except as I stumble over their antiquated stop-and-go habits. Nature has no "weeks." Quite clearly the peak traffic patterns exploited by businessmen who are eager to make the most profit in order to prove their right to live causes everybody to go in and out of the airport during two short moments in the twenty-four hours with all the main facilities shut down two-thirds of the time. All our beds around the world are empty for two-thirds of the time. Our living rooms are empty seven-eighths of the time.

The population explosion is a myth. As we industrialize, down goes the annual birth rate. If we survive, by 1985, the whole world will be industrialized, and, as with the United States, and as with all Europe and Russia and Japan today, the birth rate will be dwindling, and the bulge in population will be recognized as accounted for exclusively by those who are living longer.

When world realization of its unlimited wealth has been established there as yet will be room for the whole of humanity to stand indoors in greater New York City, with more room for each human than at an average cocktail party.

We will oscillate progressively between social concentrations in cultural centers and in multi-deployment in greater areas of our Spaceship Earth's as yet very ample accommodations. The same humans will increasingly converge for metaphysical intercourse and deploy for physical experiences.

Each of our four billion humans' shares of the Spaceship Earth's resources as yet today amount to two-hundred billion tons.

It is also to be remembered that despite the fact that you are accustomed to thinking only in dots and lines and a little bit in areas does not defeat the fact that we live in omnidirectional space-time and that a four dimensional universe provides ample individual freedoms for any contingencies.

You may very appropriately want to ask me how we are going to resolve the ever-acceleratingly dangerous impasse of world-opposed politicians and ideological dogmas. I answer, it will be resolved by the computer. Man has ever-increasing confidence in the computer; witness his unconcerned landings as airtransport passengers coming in for a landing in the combined invisibility of fog and night. While no politician or political system can ever afford to yield understandably and enthusiastically to their adversaries and opposers, all politicians can and will yield enthusiastically to the computers safe flight-controlling capabilities in bringing all of humanity in for a happy landing.

So, planners, architects, and engineers take the initiative. Go to work, and above all co-operate and don't hold back on one another or try to gain at the expense of another. Any success in such lopsidedness will be increasingly short-lived. These are the synergetic rules that evolution is employing and trying to make clear to us. They are not man-made laws. They are the infinitely accommodative laws of the intellectual integrity governing universe.

- We need to more agree upon what the objectives are. The liberal or conservative ideology.

- The objectives are not to redistribute the wealth.

- This paper has many political ideas but economics does not happen in a vacuum, it interacts w/ political and social systems.

index